T0271981

An Introduction to Intersection Homology Theory

Second Edition

An Introduction to Intersection Homology Theory

Second Edition

Frances Kirwan
Oxford University
Oxford, U.K.

Jonathan Woolf
University of Liverpool
Liverpool, U.K.

Chapman & Hall/CRC
Taylor & Francis Group
Boca Raton   London   New York

CRC Press
Taylor & Francis Group
6000 Broken Sound Parkway NW, Suite 300
Boca Raton, FL 33487-2742

© 2006 by Taylor & Francis Group, LLC
CRC Press is an imprint of Taylor & Francis Group, an Informa business

No claim to original U.S. Government works

Printed in the United States of America on acid-free paper
Version Date: 20110713

International Standard Book Number: 978-1-58488-184-1 (Hardback)

Visit the Taylor & Francis Web site at
http://www.taylorandfrancis.com

and the CRC Press Web site at
http://www.crcpress.com

Preface to the first edition

These notes are based on a course for graduate students entitled 'A beginner's guide to intersection homology theory' given in Oxford in 1987. The course was intended to be accessible to first year graduate students and to mathematicians from different areas of mathematics. The aim was to give some of the idea of the power, usefulness and beauty of intersection homology theory while only assuming fairly basic mathematical knowledge. To succeed at all in this it was necessary to give at most briefly sketched proofs of the important theorems and to concentrate on explaining the main ideas and definitions. The result is that these notes do not constitute in any sense an introductory textbook on intersection homology. Rather they are intended to be a piece of propaganda on its behalf. The hope is that mathematicians of very varied backgrounds with interests in singular spaces should find the notes readable and should be stimulated to learn in greater depth about intersection homology and use it in their work.

Over the last century ordinary homology theory for manifolds has been applied with enormous success to all sorts of different parts of mathematics. Often however ordinary homology is not as successful in dealing with problems involving singular spaces as with problems involving manifolds. In such situations it is possible that intersection homology (which coincides with ordinary homology for manifolds) may be more successful. Many examples of this phenomenon have been found since intersection homology was introduced a decade ago. It was because exactly this phenomenon has occurred in my own work in the last few years that I became an enthusiast for intersection homology, and, although by no means an expert on the subject, decided to give this course.

The goal I had in mind was to explain enough of the theory of intersection homology to be able to give a sketch (following Bernstein [15]) of the proof of Kazhdan–Lusztig conjecture (Kazhdan–Lusztig [104, 105]). This relates the representation theory of complex Lie algebras to the theory of Hecke algebras via D-modules and intersection homology, and was in fact important motivation in the development of intersection homology theory (cf. Brylinski [36]). It seemed a suitable target at which to aim, though much of the material covered on the way is just as interesting (or more so, depending on one's point of view) in its own right.

This goal influenced the structure of the second half of the course and thus the lecture notes. This first half consists of an elementary introduction to intersection homology theory. The introductory chapter, which is intended as motivation for the reader, describes three situations in which intersection homology is more successful than ordinary homology in dealing with singular spaces. The second chapter describes briefly some standard homology theory and sheaf theory; it would be helpful but not essential for the reader to be already familiar with this material. There are several different ways of defining intersection homology which vary in difficulty and elegance: Chapter 4 gives the most elementary of these and describes some of its basic properties.

The singular spaces given most attention throughout the notes are complex varieties, but intersection homology is defined for more general spaces as well (the most general being topological pseudomanifolds). The fourth chapter discusses the relationship between the intersection homology of singular complex projective varieties and an analytically defined cohomology theory, L^2-cohomology, which is a generalisation of de Rham cohomology for compact manifolds. Chapter 7 describes the important sheaf-theoretic construction and characterisations of intersection homology, due to Deligne and developed in Goresky and MacPherson [70], which imply that intersection homology is a topological invariant.

The final three chapters lead towards the proof of the Kazhdan–Lusztig conjecture which is described in Chapter 12. The tenth chapter discusses the relationship of the intersection homology with the Weil conjectures and the arithmetic of algebraic varieties defined over finite fields, while Chapter 11 describes briefly the theory of D-modules and the Riemann–Hilbert correspondence relating D-modules to intersection homology.

Nothing in these lecture notes is original work. The papers I have used most heavily are those listed in the references by Goresky and MacPherson, Borel, Bernstein, and Beilinson, Bernstein and Deligne. I would like to thank Joseph Bernstein for first suggesting several years ago that I should look at intersection homology, and all those who attended the 'beginner's guide' last year for pointing out many slips and errors. I am also grateful to Valerie Siviter for typing the original manuscript and to Terri Moss for typing the final version.

Frances Kirwan
Balliol College, Oxford
April 1988

Preface to the second edition

As a beginning graduate student trying to learn about intersection homology, I found the first edition of this book invaluable, giving, as it did, an accessible treatment with clear and simple sketches of the main ideas. Having digested it I had the confidence to go on to grapple with more specialist, technical texts and a basic framework within which to place them. Since I found it so useful, I am very pleased to be given the opportunity to co-author a second, updated edition.

This edition differs from the first in two respects. Firstly, a number of new topics have been included; some, such as Witt spaces and their bordism groups, signatures for singular spaces, perverse sheaves, and Zucker's conjecture, represent strands of thought which were omitted from the first edition, and others, such as the combinatorial construction of intersection cohomology for fans, represent subsequent developments. Secondly, some of the basic material has been revised and supplemented. The treatment of sheaf cohomology has been expanded, and given its own chapter, and more emphasis has been placed on intersection homology as a topological theory and on the rôle of generalised Poincaré duality. These changes reflect the structure and approach of a graduate course, rather unimaginatively entitled Intersection Cohomology, which I gave in Cambridge in Spring 2004.

Let me list the major revisions and supplements in more detail. The first four chapters constitute the elementary material. The introduction motivates the subject by giving examples of the utility of intersection homology. The old second chapter has been split into two, the first part reviewing simplicial and singular homology, and the second reviewing sheaf cohomology from both the Čech and derived functor viewpoints. The latter contains new material on derived categories of sheaves, which are a fundamental technical tool. The treatment of intersection homology in the fourth chapter has been revised and expanded to apply to pseudomanifolds rather than just to complex projective varieties. The latter now appear as a nice class of examples with especially good properties.

Rational intersection homology satisfies generalised Poincaré duality for a class of singular spaces called Witt spaces. These include all pseudomanifolds with only even dimensional strata, such as complex projective varieties, and also those pseudomanifolds satisfying a certain condition on the links of any odd codimensional strata. It is possible to define a bordism invariant signature for a Witt space. Chapter 5 discusses this material (which was not in the first edition). It culminates in a sketch of Siegel's beautiful computation [162] equating the bordism groups of $4n$-dimensional Witt spaces with the Witt group of symmetric rational bilinear forms.

Chapter 6 explains the relation of intersection cohomology to the analytically defined L^2-cohomology. It now contains a (very) brief introduction to the Hodge theory of L^2-cohomology and a new section, based on Zucker [184], on locally symmetric varieties and Zucker's conjecture.

Chapter 7 explains how the intersection homology groups can be obtained as the (hyper)cohomology of an intersection sheaf complex. This complex can be axiomatically characterised independently of the stratification, leading to a proof of intersection homology's topological invariance. It also has a new section on constructible sheaves and Verdier duality. This duality, a contravariant equivalence on the constructible derived category of sheaves, plays a fundamental rôle in intersection homology theory.

There is a beautiful Abelian subcategory, the perverse sheaves, which is preserved by Verdier duality and whose simple objects are intersection sheaf complexes, possibly with twisted coefficients, supported on the strata. Chapter 8 gives a simple introduction to this deep theory. The nearby and vanishing cycles of a fibre of a complex analytic map are introduced as important examples of perverse sheaves. An amplified section on Beilinson, Bernstein, Deligne and Gabber's decomposition theorem completes the chapter.

The new Chapter 9 provides an elementary treatment of the combinatorial intersection cohomology of a fan. When the fan is rational there is a corresponding toric variety whose intersection cohomology agrees with this combinatorial invariant of the fan. In this situation, deep results, such as the decomposition theorem, have relatively simple combinatorial proofs. The chapter ends with a discussion of Stanley's conjectures on the generalised h-vector of a fan.

The discussion of the Weil conjectures, $\mathcal{D}$-modules, the Riemann–Hilbert correspondence and the Kazhdan–Lusztig conjecture in Chapters 9, 10 and 11 is virtually unchanged, apart from some corrections, in particular to the definition of étale cohomology.

I have tried to write in the spirit of the first edition, maintaining the book as an introductory guide, or even a piece of propaganda on behalf of the subject, rather than a textbook. This means that many results are quoted, or presented with only a sketch proof. In order that the interested reader can delve further I have attempted to provide a comprehensive bibliography. Each chapter concludes with a brief section suggesting further reading. Nevertheless, intersection homology is a large and growing subject, touching on many aspects of topology, geometry and algebra and with a correspondingly large research literature. I will undoubtedly have made omissions and oversights, for which I can only apologise. One topic which is prominent by its absence is Saito's theory of mixed Hodge modules and the existence of a Hodge structure on the intersection cohomology of a complex projective variety.

None of the results in this book are original, and I owe many debts to the clear expositions in the references, whilst accepting full responsibility for any errors.

Most of this second edition was written during my time at Christ's College, Cambridge and I am very grateful for their financial, social and culinary support. I would like to thank the students who sat through my course and remained cheerful until the end. I am also very grateful to Aaron Lauda for LaTeX-ing the original manuscript of the first edition and to Ivan Smith for his indefatigable proof-reading and numerous helpful comments and suggestions (though again, the remaining errors are mine).

Special thanks go to Soumhya for her patience and encouragement, particularly during my more irascible moments. Finally, I wish to thank Frances Kirwan for her invaluable help during the writing of this second edition and for introducing me to intersection homology and infecting me with her enthusiasm for the subject.

Jonathan Woolf

University of Liverpool

October 2005

List of Figures

Contents

Chapter 1

Introduction

Homology theory was introduced by Poincaré just over a century ago in order to study the topology of manifolds. As he foresaw, it has been of immense importance in many areas of mathematics including algebraic and differential geometry, differential equations and group theory.

One of Poincaré's principal motivations was to study the intersection theory of submanifolds. The key result in this respect is Poincaré duality: this is the existence of a non-degenerate pairing between the rational homology in dimensions i and j of a closed oriented manifold M, where $i + j = \dim M$. In terms of intersection theory, a closed oriented i-dimensional submanifold generates an i-dimensional homology class and for submanifolds A and B with $\dim A + \dim B = \dim M$, and which are in 'general position', the pairing of the corresponding classes is given by counting the points of $A \cap B$ (with appropriate signs).

The close relationship between the homology of a manifold and the intersection theory of its submanifolds is very important. However, it is not the only reason why homology and its dual theory cohomology are powerful invariants for manifolds. Much of their importance comes from the fact that they can be interpreted in new ways, and often imbued with extra structure, when the underlying manifold itself has extra structure. Important examples of this are

(a) the homology of a smooth manifold can be interpreted in terms of the critical points of a generic smooth function via Morse theory;

(b) the real cohomology of a smooth manifold can be interpreted in terms of differential forms via the de Rham isomorphism;

(c) the Hodge theorem tells us that the real cohomology of a compact Riemannian manifold, i.e. a smooth manifold with a metric, is isomorphic to the space of harmonic forms. Put another way, each de Rham cohomology class has a unique 'minimal energy' representative;

1

(d) the complex cohomology of a compact Kähler manifold, i.e. a compact Riemannian manifold with a compatible complex structure, has a Hodge decomposition reflecting the classification of harmonic forms by the degree of their holomorphic and anti-holomorphic components;

(e) the geometry of intersections with a generic hyperplane section provides information about the homology of a non-singular complex projective variety. This is encoded in the Lefschetz hyperplane and hard Lefschetz theorems.

Unfortunately, for non-manifolds the close relationship between homology and intersection theory breaks down and, in particular, Poincaré duality fails. Furthermore, we do not have analogues of the rich set of interpretations and extra structure (a)–(e) above, not even for the (co)homology of a singular complex projective variety.

In the 1970s a new sort of homology, called intersection homology, was introduced by Goresky and MacPherson. Many others have helped to develop its theory since then. Intersection homology coincides with ordinary homology for manifolds but often has 'better' properties than homology for singular spaces. In particular, and this was Goresky and MacPherson's motivation, intersection homology satisfies an analogue of Poincaré duality for a wide class of singular spaces. It is a minor mathematical miracle that intersection homology also has (at least in part) analogues of (a)–(e) above. The remainder of this introductory chapter is given over to a slightly more detailed discussion of this phenomenon with the aim of whetting the reader's appetite before the definition of intersection homology is given.

To avoid long-winded definitions of the classes of singular spaces involved we will restrict our attention to complex projective varieties (in any case these form the most interesting class of examples). A **complex projective variety** X is a subset of a complex projective space

$$\mathbb{CP}^m = \frac{\mathbb{C}^{m+1} - \{0\}}{\mathbb{C} - \{0\}} = \{\text{complex lines in } \mathbb{C}^{m+1}\},$$

and is defined by the vanishing of homogeneous polynomials. Let us write

$$\left(x_0 : x_1 : \ldots : x_m\right) \tag{1.1}$$

for the complex line in $\mathbb{C}^{m+1}$ spanned by a non-zero vector $(x_0, \ldots, x_m)$ in $\mathbb{C}^{m+1}$. Then X is of the form

$$X = \left\{(x_0 : \ldots : x_m) \in \mathbb{CP}^m \mid f_j(x_0, \ldots x_m) = 0, \quad 1 \le j \le M\right\} \tag{1.2}$$

where $f_1, \ldots, f_M$ are homogeneous polynomials in $m+1$ variables. The homogeneity of f_j implies that the condition

$$f_j\left(x_0, \ldots, x_m\right) = 0$$

is independent of the choice of vector $(x_0, \ldots, x_m) \in \mathbb{C}^{m+1} - \{0\}$ representing the point $(x_0 : x_1 : \cdots : x_m)$ of $\mathbb{CP}^m$.

Complex projective space $\mathbb{CP}^m$ is a complex manifold with local coordinates

$$(x_0 : \ldots : x_m) \rightarrow \left(\frac{x_0}{x_j}, \ldots, \frac{x_{j-1}}{x_j}, \frac{x_{j+1}}{x_j}, \ldots \frac{x_m}{x_j} \right)$$

identifying the open subsets

$$\{(x_0 : \ldots : x_m) \in \mathbb{CP}^m \mid x_j \neq 0\}, \qquad 0 \leq j \leq m$$

of $\mathbb{CP}^m$ with $\mathbb{C}^m$. We say that X is **non-singular** if, locally, we can choose $f_1, \ldots, f_M$ in (1.2) so that the Jacobian matrix $\left(\frac{\partial f_i}{\partial x_j} \right)$ has rank M. In this case X becomes a complex submanifold of $\mathbb{CP}^m$.

We denote the homology and cohomology (with coefficients in $\mathbb{C}$) of a space X by

$$H_*(X) \quad \text{and} \quad H^*(X)$$

respectively (see Chapter 2 for the definitions). They are graded complex vector spaces; the cohomology is simply the vector space dual of the homology. Both homology and cohomology are topological invariants.

1.1 Poincaré duality

Suppose X is a compact oriented manifold. Poincaré duality states that there are non-degenerate bilinear pairings

$$H_i(X) \otimes H_{n-i}(X) \rightarrow \mathbb{C} \qquad (1.3)$$

for each $0 \leq i \leq n$ where $n = \dim_\mathbb{R} X$. In particular

$$\dim_\mathbb{C} H_i(X) = \dim_\mathbb{C} H_{n-i}(X).$$

As mentioned above these pairings have a geometric interpretation in terms of intersections of submanifolds.

Poincaré duality does not hold for the cohomology of a singular projective variety X. Here is a simple example. First recall that the complex projective line $\mathbb{CP}^1$ can be identified with the extended complex plane $\mathbb{C} \cup \{\infty\}$. Topologically it is a 2-dimensional sphere. Thus

$$H_i(\mathbb{CP}^1) = \begin{cases} \mathbb{C} & i = 0, 2 \\ 0 & \text{otherwise} \end{cases}$$

(Spanier, [163, Ch. 4, §6, Thm. 6], or Example 2.1.5 below). Now let X be the complex projective variety

$$\{(x : y : z) \in \mathbb{CP}^2 \mid yz = 0\}. \qquad (1.4)$$

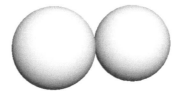

Figure 1.1: Topological picture of the projective curve $yz = 0$ in $\mathbb{CP}^2$.

Then X is the union of the two subsets

$$\{(x:y:z) \in \mathbb{CP}^2 \mid y = 0\} \quad \text{and} \quad \{(x:y:z) \in \mathbb{CP}^2 \mid z = 0\}$$

of $\mathbb{CP}^2$. These subsets are each homeomorphic to $\mathbb{CP}^1$ and meet in the single point $(1:0:0)$, see Figure 1.1. It can easily be shown, for example by using the Mayer–Vietoris sequence (Spanier [163, Ch. 4, §6]), that

$$H_i(X) = \begin{cases} \mathbb{C} & i = 0 \\ 0 & i = 1 \\ \mathbb{C} \oplus \mathbb{C} & i = 2. \end{cases}$$

Hence Poincaré duality cannot hold in this case. One way to remedy this is by introducing the **intersection homology groups** $IH_*(X)$ of X. These are complex vector spaces which are topological invariants of X. They have the property that for any complex projective variety of *complex* dimension n, whether singular or not, there are non-degenerate pairings

$$IH_i(X) \otimes IH_{2n-i}(X) \to \mathbb{C}$$

for $0 \leq i \leq 2n$. For non-singular X there is a natural isomorphism

$$IH_*(X) \cong H_*(X)$$

and this pairing is identified with that in (1.3). The existence of the pairings for intersection homology is a topological fact; it does not rely on the complex geometry of X.

1.2   Morse theory for singular spaces

Suppose X is a compact smooth manifold. A smooth function $f \colon X \to \mathbb{R}$ is called a **Morse function** (Milnor [135]) if at each point x of the set

$$C(f) = \{x \in X \mid df(x) = 0\}$$

of critical points of f the Hessian $H_x(f)$ is non-degenerate (in which case the critical points are isolated and hence finite in number). Here $H_x(f)$ is the bilinear form on T_xX given in local coordinates $x_1, \ldots, x_m$ by the matrix

$$\left(\frac{\partial^2 f}{\partial x_i \partial x_j} \right)$$

of the second partial derivatives of f at x. We shall also require for simplicity of notation that if x and y are distinct critical points then $f(x) \neq f(y)$.

The set of Morse functions is open and dense in the set of all smooth functions on X.

Proposition 1.2.1. *If $f: X \to \mathbb{R}$ is a Morse function then for each $y \in \mathbb{R}$ either*

(i) *$y \neq f(x)$ for all critical points $x \in C(f)$, in which case if $\varepsilon > 0$ is small enough the map*

$$H_k(X_{y-\varepsilon}) \to H_k(X_{y+\varepsilon})$$

induced by the inclusion of open sets

$$X_{y-\varepsilon} = \{x \in X | f(x) < y - \varepsilon\} \hookrightarrow X_{y+\varepsilon} = \{x \in X | f(x) < y + \varepsilon\}$$

is an isomorphism for all k; or

(ii) *$y = f(x)$ for some (unique) critical point $x \in C(f)$, in which case there is an integer $I(f;x)$ such that if $\varepsilon > 0$ is small enough the map*

$$H_k(X_{y-\varepsilon}) \to H_k(X_{y+\varepsilon})$$

induced by the inclusion is an isomorphism except when k is $I(f;x)$ or $I(f;x) - 1$, and for these values of k it fits into an exact sequence

$$0 \to H_{I(f;x)}(X_{y-\varepsilon}) \to H_{I(f;x)}(X_{y+\varepsilon}) \to \mathbb{C}$$
$$\to H_{I(f;x)-1}(X_{y-\varepsilon}) \to H_{I(f;x)-1}(X_{y+\varepsilon}) \to 0.$$

Another way to express this is to say that the relative homology is given by

$$H_k(X_{y+\varepsilon}, X_{y-\varepsilon}) = \begin{cases} 0 & \text{if } k \neq I(f;x) \\ \mathbb{C} & \text{if } k = I(f;x). \end{cases}$$

The integer $I(f;x)$ is called the **Morse index** of the critical point x for the function f. It is the number of negative eigenvalues of the Hessian $H_x(f)$.

As a consequence of Proposition 1.2.1 we obtain the famous **Morse inequalities**, which are most easily written in the following form

$$\sum_{x \in C(f)} t^{I(f;x)} - \sum_{i \geq 0} t^i \dim_{\mathbb{C}} H_i(X) = (1+t)R(t) \qquad (1.5)$$

where $R(t)$ is a polynomial in t with *non-negative* integer coefficients. In particular this implies that the dimension of $H_i(X)$ is at most the number of $x \in C(f)$ with $I(f;x) = i$, but the Morse inequalities contain stronger information than this. For example if $I(f;x)$ is even for all $x \in C(f)$ the Morse inequalities can only work if $R(t) = 0$, i.e. if the dimension of $H_i(X)$ is equal to the number of $x \in C(f)$ with $I(f;x) = i$ for all i.

Morse theory can be generalised to the case when X is a singular projective variety provided that intersection homology is used instead of ordinary homology.

Proposition 1.2.2. *Suppose that X is a complex projective variety. The set of all functions $f: X \to \mathbb{R}$ which extend to smooth functions on projective space contains a dense open subset such that for any $f: X \to \mathbb{R}$ in this subset there exists a finite set $C(f) \subseteq X$ with the following properties.*

(i) *If $y \in \mathbb{R} - \{f(x) \mid x \in C(f)\}$ then there is an isomorphism*

$$IH_k(X_{y-\varepsilon}) \cong IH_k(X_{y+\varepsilon})$$

for all sufficiently small $\varepsilon > 0$. The isomorphism is induced by the inclusion.

(ii) *If $y = f(x)$ for some $x \in C(f)$ then this x is unique, and there exists an integer $I(f;x) \geq 0$, called the Morse index of x for f and a complex vector space A_x such that if $\varepsilon > 0$ is small enough then*

$$IH_k(X_{y-\varepsilon}) \cong IH_k(X_{y+\varepsilon})$$

unless k is $I(f;x)$ or $I(f;x) - 1$, and there is an exact sequence

$$0 \to IH_{I(f;x)}(X_{y-\varepsilon}) \to IH_{I(f;x)}(X_{y+\varepsilon}) \to A_x$$
$$\to IH_{I(f;x)-1}(X_{y-\varepsilon}) \to IH_{I(f;x)-1}(X_{y+\varepsilon}) \to 0.$$

Here A_x depends on x and X but *not* on the function f. In fact A_x is determined by the singularity of X at x. If x is a non-singular point of X then $A_x = \mathbb{C}$. (See Goresky and MacPherson [74, 71] for more details.)

Warning 1. Morse theory in the sense of Proposition 1.2.1 applies to a Morse function on a compact smooth manifold. However, the above generalisation to singular spaces, in particular the existence of a Morse index for intersection homology, relies in an essential way on the complex geometry of X as well as on the properties of intersection homology.

From Proposition 1.2.2 one gets **generalised Morse inequalities**

$$\sum_{x \in C(f)} t^{I(f;x)} \dim_{\mathbb{C}} A_x - \sum_{i \geq 0} t^i \dim_{\mathbb{C}} IH_i(X) = (1+t)Q(t) \qquad (1.6)$$

where $Q(t)$ is a polynomial in t with non-negative integer coefficients.

Remark 1.2.3. If X is a singular projective variety there does not in general exist a Morse index for ordinary homology. As y moves through a critical value the homology may change in a whole range of dimensions (Goresky and MacPherson [71]).

1.3   de Rham cohomology and L^2-cohomology

When X is a manifold the cohomology $H^*(X)$ can be identified with the de Rham cohomology $H_{\mathrm{dR}}^*(X)$ which is defined as follows (see Bott and Tu [26] for more details).

Let TX be the tangent bundle to X and let T^*X be the cotangent bundle. A differential r-form ω on X is a C^∞ section of the r-fold exterior product $\wedge^r T^*X$ of the cotangent bundle. In (real) local coordinates $y_1, \ldots, y_m$ we have

$$\omega(y) = \sum_{i_1 < \ldots < i_r} a_{i_1 \ldots i_r}(y) \mathrm{d}y_{i_1} \wedge \ldots \wedge \mathrm{d}y_{i_r}$$

where each $a_{i_1 \ldots i_r}$ is a smooth real-valued function of $y = (y_1, \ldots, y_m)$.

Let $A^r(X; \mathbb{R})$ be the space of all differential r-forms and let

$$A^r(X) = A^r(X; \mathbb{R}) \otimes_{\mathbb{R}} \mathbb{C}$$

be the space of all complex valued differential r-forms. There is a map

$$\mathrm{d} \colon A^r(X) \to A^{r+1}(X)$$

defined in local coordinates by

$$\mathrm{d}\omega = \sum_{i_1 < \ldots < i_r} \sum_j \left(\frac{\partial a_{i_1 \ldots i_r}}{\partial y_j} \right) dy_j \wedge dy_{i_1} \wedge \ldots \wedge dy_{i_r}$$

when ω is as above.

Then $d^2 = 0$ (by the symmetry of the second partial derivative of a C^∞ function). The rth de Rham complex cohomology group of X is by definition the quotient group

$$H_{\mathrm{dR}}^r(X) = \frac{\ker d \colon A^r(X) \to A^{r+1}(X)}{\operatorname{im} d \colon A^{r-1}(X) \to A^r(X)} \tag{1.7}$$

Theorem 1.3.1. *(de Rham theorem, see e.g. Griffiths and Harris [77, p43]).* $H_{\mathrm{dR}}^r(X)$ *is canonically isomorphic to $H^r(X)$.*

Together with the famous Hodge theorem (Griffiths and Harris [77, Ch. 0, §6]) this can be used to give $H^r(X)$ extra structure when X is a Kähler manifold, that is when X has a complex structure and a compatible metric. The Hodge theorem implies that every de Rham cohomology class in $H_{\mathrm{dR}}^r(X)$ contains a unique harmonic differential r-form ω which can be written uniquely as a sum of harmonic (p, q)-forms where $p + q = r$. A (p, q)-form is one which can be written locally with respect to *complex* local coordinates $z_1, \ldots, z_n$ as a sum of terms of the form

$$\alpha \, dz_{i_1} \wedge \ldots \wedge dz_{i_p} \wedge d\bar{z}_{j_1} \wedge \ldots \wedge d\bar{z}_{j_q}$$

where $i_1 < \ldots < i_p$ and $j_1 < \ldots < j_q$ and α is a smooth function.

Theorem 1.3.2 (Hodge decomposition). *Suppose X is a Kähler manifold. We can write*

$$H^i(X) = \bigoplus_{p+q=i} H^{p,q} \tag{1.8}$$

where $H^{p,q}$ is the space of harmonic (p, q)-forms. It is a complex subspace of $H^i(X)$ and

$$H^{p,q} = \overline{H^{q,p}}.$$

Note that for complex conjugation to make sense we need a real structure on $H^i(X)$, i.e. a real subspace V of $H^i(X)$ such that

$$H^i(X) = V \otimes_{\mathbb{R}} \mathbb{C}.$$

We take V to be the ith cohomology group $H^i(X; \mathbb{R})$, with real coefficients, of X.

The Hodge decomposition implies that if i is *odd* then

$$\dim_{\mathbb{C}} H^i(X) = 2 \sum_{p<q, p+q=i} \dim_{\mathbb{C}} H^{p,q}$$

is *even*.

A non-singular complex projective variety is a Kähler manifold with the Kähler metric given by the restriction of the Fubini–Study metric on complex projective space (Griffiths and Harris [77, p31]). Thus the cohomology of a non-singular projective variety has a Hodge decomposition. This is very useful for studying non-singular projective varieties. If one allows X to vary in a holomorphic way depending on some continuous parameters then $H^i(X)$ is essentially independent of X but the Hodge filtration

$$H^i(X) = F^0 \supseteq F^1 \supseteq \ldots \supseteq F^i$$

of $H^i(X)$ defined by

$$F^p = \bigoplus_{j \geq p} H^{j, i-j}$$

varies holomorphically with X in an interesting way. This leads to Griffiths' theory of the variation of Hodge structure which gives information about moduli spaces (Griffiths [76]).

Figure 1.2: Topological picture of the projective curve $x^3 + y^3 = xyz$ in $\mathbb{CP}^2$.

The cohomology of a singular complex projective variety need not have a Hodge decomposition in the above sense. Here is a simple example. Let X be the complex projective variety

$$\left\{ (x : y : z) \in \mathbb{CP}^2 \mid x^3 + y^3 = xyz \right\}. \tag{1.9}$$

Then it is not hard to check that topologically X is a 2-dimensional sphere with two points identified, see Figure 1.2. Thus

$$H^i(X) = \begin{cases} \mathbb{C} & i = 0 \\ \mathbb{C} & i = 1 \\ \mathbb{C} & i = 2. \end{cases}$$

In particular $\dim_{\mathbb{C}} H^1(X)$ is odd so there cannot be a Hodge decomposition of the cohomology of X.

One would like to have some sort of analytically defined cohomology when X is singular (at least when X is a singular complex projective variety, perhaps in more general cases too) analogous to de Rham cohomology and canonically isomorphic to intersection cohomology. With luck this could then be used to give analytical proofs that intersection cohomology has a Hodge decomposition and satisfies Poincaré duality and the hard Lefschetz theorem (see Theorem 1.4.2). It should have all sorts of other spin-offs as well, just as the de Rham theorem does.

It is conjectured (and proven in some cases) that there is such a cohomology theory defined analytically (see Cheeger, Goresky and MacPherson [48]). It is called L^2-cohomology and is defined as follows.

Let $X \subseteq \mathbb{CP}^m$ be a projective variety of complex dimension n. Let Σ be the set of singular points of X. The restriction of the Fubini–Study metric on $\mathbb{CP}^m$ to $X - \Sigma$ gives us a Riemannian metric on the manifold $X - \Sigma$, i.e. an inner product g_x on each tangent space $T_x(X - \Sigma) = T_x X$ which varies smoothly with $x \in X - \Sigma$. This inner product g_x induces inner products on the cotangent space $T^*_x(X - \Sigma)$ and its exterior powers $\Lambda^i T^*_x(X - \Sigma)$ for all $i > 0$ and $x \in X - \Sigma$.

Given a smooth i-form ω on $X - \Sigma$ we have a smooth function $\|\omega\|^2$ on $X - \Sigma$ defined by $x \to \|\omega(x)\|^2$ where $\| \ \|$ is the norm on $\Lambda^i T^*_x(X - \Sigma)$ induced by the inner product. The i-form is called **square integrable** if this function $\|\omega\|^2$ is integrable over $X - \Sigma$ with respect to the volume form induced by the metric and the natural orientation on $X - \Sigma$. (For more details on differential geometry see e.g. Spivak [164], Sternberg [170], Warner [175].)

Let $L^i(X - \Sigma) \subseteq A^i(X - \Sigma)$ be the space of square-integrable differential i-forms on $X - \Sigma$. The L^2-cohomology of $X - \Sigma$ is defined to be

$$H^i_{(2)}(X - \Sigma) = \frac{\{\omega \in L^i(X - \Sigma) | d\omega = 0\}}{\{\eta \in L^i(X - \Sigma) \mid \exists \zeta \in L^{i-1}(X - \Sigma), \ d\zeta = \eta\}}. \qquad (1.10)$$

Note that d may not map $L^{i-1}(X - \Sigma)$ into $L^i(X - \Sigma)$.

Of course if X is non-singular then $X - \Sigma = X$ is compact so $L^i(X - \Sigma)$ is $A^i(X)$ for all i and

$$H^i_{(2)}(X - \Sigma) = H^i_{\mathrm{dR}}(X).$$

Conjecture 1.3.3. *(See Cheeger, Goresky and MacPherson [48]). If X is a singular projective variety then $H^*_{(2)}(X - \Sigma)$ is isomorphic to $IH^*(X)$.*

It is not even known that $H^i_{(2)}(X - \Sigma)$ is finite-dimensional in general, but the conjecture is known to be true when X has isolated conical singularities or is a locally symmetric variety (see Chapter 6).

The theory of L^2 harmonic forms on $X - \Sigma$ is difficult and not well understood. In particular, it is not known if an analogue of the Hodge decomposition holds for L^2-cohomology. However the intersection cohomology of a complex projective variety can be given a Hodge decomposition (without the interpretation of $H^{p,q}$ as the space of harmonic (p,q)-forms) using other techniques. Sadly, these are beyond the scope of this book — see Saito [150] for more information.

1.4   The cohomology of projective varieties

Suppose X is a non-singular complex projective variety of complex dimension n in $\mathbb{CP}^m$. A hyperplane $H \subset \mathbb{CP}^m$ is a linear subspace of complex codimension one or, equivalently, a complex projective variety defined by the vanishing of a single linear polynomial. By considering how a generic hyperplane intersects X we can obtain information about the cohomology of X. ("Generic" means that the property we are interested in will not necessarily hold for every hyperplane but it will hold for most – more precisely for those in a dense open subset of the space of all possible linear equations.)

Theorem 1.4.1 (Lefschetz hyperplane theorem). *Let $H \subset \mathbb{CP}^m$ be a generic hyperplane. Then the restriction map*

$$H^i(X) \to H^i(X \cap H)$$

is an isomorphism for $i < n - 1$ and is an injection for $i = n - 1$.

There is a hyperplane class $[H] \in H^2(X)$ which is independent of choice of hyperplane. (It is the 'Poincaré dual' of the intersection $X \cap H$.) Cohomology has a product structure and multiplication by the hyperplane class defines a map

$$L : H^i(X) \to H^{i+2}(X).$$

Theorem 1.4.2 (Hard Lefschetz theorem). *Multiplication by powers of the hyperplane class defines isomorphisms*

$$L^i \colon H^{n-i}(X) \to H^{n+i}(X).$$

In particular the map

$$L \colon H^k(X) \to H^{k+2}(X)$$

is injective if $k < n$, so that $\dim_{\mathbb{C}} H^k(X) \leq \dim_{\mathbb{C}} H^{k+2}(X)$, and surjective if $k + 2 > n$, so that $\dim_{\mathbb{C}} H^k(X) \geq \dim_{\mathbb{C}} H^{k+2}(X)$.

The hard Lefschetz theorem enables us to refine the Hodge decomposition in the following way. If $p + q = n - i$ where $0 \leq i \leq n$ let

$$H^{p,q}_{\text{prim}} = \left\{ \xi \in H^{p,q} \mid L^{i+1}(\xi) = 0 \right\}.$$

Here "prim" stands for primitive cohomology. Then if $p + q \leq n$ we have

$$H^{p,q} = H^{p,q}_{\text{prim}} \oplus L\left(H^{p-1,q-1}_{\text{prim}}\right) \oplus L^2\left(H^{p-2,q-2}_{\text{prim}}\right) \oplus \cdots.$$

This refinement is compatible with Poincaré duality in the following sense.

Theorem 1.4.3 (Hodge signature theorem). *Let p and q be integers between 0 and n, and suppose that $\xi \in H^{p,q}_{\text{prim}}(X)$ is non-zero. Then under the Poincaré duality pairing*

$$H^{p,q}(X) \otimes H^{2n-p-q}(X) \to \mathbb{C}$$

the pairing of $\xi \in H^{p+q}(X)$ with the element

$$\left(\sqrt{-1}\right)^{p-q} (-1)^{(n-p-q)(n-p-q-1)/2} L^{n-p-q}(\xi) \in H^{2n-p-q}(X)$$

is a strictly positive real number.

Proofs of the Lefschetz hyperplane, hard Lefschetz and Hodge signature theorems can be found in Griffiths and Harris [77, Ch. 0, §7 and Ch. 1 §2].

These three theorems do not apply to the cohomology of singular complex projective varieties. To see that the Lefschetz hyperplane theorem can fail we consider the variety X in $\mathbb{CP}^4$ defined by the four equations

$$x_i x_j = 0 \qquad \text{for } i \in \{0, 1\} \text{ and } j \in \{3, 4\}.$$

It is the union of the two copies $\{x_0 = x_1 = 0\}$ and $\{x_3 = x_4 = 0\}$ of $\mathbb{CP}^2$ which meet in the single point $(0 : 0 : 1 : 0 : 0)$. We can compute

$$H^i(X) = \begin{cases} \mathbb{C} & i = 0 \\ 0 & i = 1 \\ \mathbb{C} \oplus \mathbb{C} & i = 2 \\ 0 & i = 3 \\ \mathbb{C} \oplus \mathbb{C} & i = 4 \end{cases}$$

For a generic hyperplane $H \subset \mathbb{CP}^4$ the intersection $X \cap H$ is the disjoint union of two copies of $\mathbb{CP}^1$ so that

$$H^i(X \cap H) = \begin{cases} \mathbb{C} \oplus \mathbb{C} & i = 0 \\ 0 & i = 1 \\ \mathbb{C} \oplus \mathbb{C} & i = 2 \end{cases}$$

In particular $\dim_\mathbb{C} H^0(X \cap H) \neq \dim_\mathbb{C} H^0(X)$ and the Lefschetz hyperplane theorem does not hold. The hard Lefschetz theorem also fails because $\dim_\mathbb{C} H^0(X) \neq \dim_\mathbb{C} H^4(X)$. Finally, since we do not necessarily have either Poincaré duality or a Hodge decomposition for the cohomology of a singular variety, we cannot even make sense of the statement of the Hodge signature theorem.

Once more, intersection cohomology remedies the situation. Appropriate analogues of the Lefschetz hyperplane, hard Lefschetz and Hodge signature theorems all hold for the intersection cohomology of a singular projective variety.

Often Poincaré duality, the Hodge decomposition, Lefschetz hyperplane, hard Lefschetz and Hodge decomposition theorems for the cohomology of a non-singular complex projective variety are known collectively as the 'Kähler package'. Together they give us a powerful set of tools for studying non-singular varieties. It is amazing, and rather wonderful, that a similar package exists for the intersection cohomology of singular varieties. Sadly, apart from Poincaré duality, we will not prove the Kähler package for intersection cohomology in this book as the proofs are quite demanding. However, we will indicate the sort of methods required in §4.10.

This ends the introduction. Its aim was to make the reader sufficiently interested in intersection homology to want to find out how it is actually defined. In order to do this we first need to review some ordinary (co)homology theory and sheaf theory. Before we gallop off on this task we pause to note that this book, with its emphasis on introducing the reader to the various interesting aspects of intersection homology theory, is rather ahistorical. Readers are therefore strongly encouraged to read Kleiman's beautiful history [113] of the subject as a corrective.

Chapter 2

Review of homology and cohomology

If X is a compact manifold there are several ways of defining the homology and cohomology groups $H_i(X)$ and $H^i(X)$ of X which all lead to essentially the same thing in the end: simplicial homology and cohomology; singular homology and cohomology; Čech cohomology of sheaves; sheaf cohomology via derived functors and de Rham cohomology. We shall review the first two of these in this chapter and the second two in the next; for the definition of de Rham cohomology see Section 1.3.

We fix a field $\mathbb{F}$ (which the reader is welcome to assume is $\mathbb{Q}$, $\mathbb{R}$ or $\mathbb{C}$).

2.1 Simplicial homology

Simplicial homology is the most prosaic and least elegant definiton. It is useful for working out examples.

Definition 2.1.1. An n-**simplex** σ in $\mathbb{R}^N$ is the convex hull of a set of points $v_0, \ldots, v_n$ such that $v_1 - v_0, v_2 - v_0, \ldots, v_n - v_0$ are n linearly independent vectors in $\mathbb{R}^N$. Then $v_0, \ldots, v_n$ are the **vertices** of σ and n is the dimension of σ. The **faces** of σ are the $(n-1)$-simplices whose vertices are also vertices of σ, for example the convex hull of $v_0, v_2, v_3, \ldots, v_n$.

An **orientation** of an n-simplex σ is an ordering of its vertices determined up to even permutation.

Definition 2.1.2. A **simplicial complex** in $\mathbb{R}^n$ is a set N of simplices such that

(i) if $\sigma \in N$ then every face of σ is in N;

(ii) if $\sigma, \tau \in N$ and $\sigma \cap \tau \neq \emptyset$ then $\sigma \cap \tau$ is a simplex whose vertices are also vertices of both σ and τ;

(iii) if $x \in \sigma \in N$ then there is a neighbourhood $\mathcal{U}$ of x in $\mathbb{R}^N$ such that $\mathcal{U} \cap \tau \neq \emptyset$ for only finitely many simplices $\tau \in N$.

Definition 2.1.3. The **support**

$$|N| = \bigcup_{\sigma \in N} \sigma$$

of a simplicial complex N in $\mathbb{R}^N$ is the union of the simplices which belong to it. A **triangulation** of a topological space X is a homeomorphism $T \colon |N| \to X$ where N is a simplicial complex.

We shall assume henceforth that X is triangulable, i.e. that X has a triangulation $T \colon |N| \to X$. Note that N is finite if and only if X is compact. For each $\sigma \in N$ choose an orientation of σ. Let

$$N^{(i)} = \left\{ \sigma \in N \,\middle|\, \sigma \text{ is an } i\text{-simplex} \right\}.$$

An i-chain of N with coefficients in a field $\mathbb{F}$ is a formal linear combination

$$\xi = \sum_{\sigma \in N^{(i)}} \xi_\sigma \sigma$$

where the coefficients ξ_σ are elements of $\mathbb{F}$ and only finitely many of them are non-zero. The space $C_i(N)$ of i-chains in N is a vector space over $\mathbb{F}$ with basis $N^{(i)}$. The boundary map

$$\partial \colon C_i(N) \to C_{i-1}(N)$$

is the unique linear map such that if $\sigma \in N^{(i)}$ then

$$\partial \sigma = \sum_{\tau \text{ face of } \sigma} \pm \tau$$

where the sign $\pm$ is 1 if the chosen orientation on τ is obtained from the chosen orientation, $v_0, \ldots, v_i$ say, on σ by omitting some v_j where j is even, and is -1 otherwise. Then

$$\partial^2 \colon C_i(N) \to C_{i-2}(N)$$

is 0, i.e.

$$\mathrm{im}\, (\partial \colon C_i(N) \to C_{i-1}(N)) \subseteq \ker\, (\partial \colon C_{i-1}(N) \to C_{i-2}(N)).$$

Definition 2.1.4. The ith **homology group** of N with coefficients in $\mathbb{F}$ is

$$H_i(N) = \frac{\ker \partial \colon C_i(N) \to C_{i-1}(N)}{\mathrm{im}\ \partial \colon C_{i+1}(N) \to C_i(N)}.$$

If $T\colon |N| \to X$ is a triangulation of X then we define the ith homology group of X with respect to T as

$$H_i^T(X) = H_i(N).$$

We also write $C_i^T(X)$ for $C_i(N)$.

In fact $H_i^T(X)$ does not depend on the triangulation T chosen (see Theorem 2.2.1 below). It is a definition of homology which is usually easy to calculate in examples.

Example 2.1.5. Let X be the 2-dimensional sphere S^2 and let $T\colon |N| \to X$ be the triangulation indicated by the diagram.

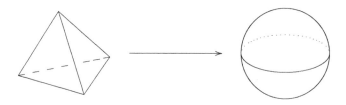

Then im $\partial\colon C_1(N) \to C_0(N)$ is spanned by $\{v_i - v_j \,|\, 0 \le i < j \le 3\}$ so $H_0^T(X) = \mathbb{C}$. Also

$$\ker \partial\colon C_1(N) \to C_0(N) = \text{im } \partial\colon C_2(N) \to C_1(N)$$

so $H_1^T(X) = 0$. Finally $\ker \partial\colon C_2(N) \to C_1(N)$ is spanned by

$$(v_0 v_1 v_2) - (v_0 v_1 v_3) + (v_0 v_2 v_3) - (v_1 v_2 v_3)$$

so $H_2^T(X) = \mathbb{F}$.

Definition 2.1.6. A triangulation $T\colon |N| \to X$ is a **refinement** of a triangulation $\tilde{T}\colon |\tilde{N}| \to X$ if for each $\sigma \in N$ there exists some $\tilde{\sigma} \in \tilde{N}$ such that $T(\sigma) \subseteq \tilde{T}(\tilde{\sigma})$.

If T is a refinement of $\tilde{T}$ then there is a natural map

$$C_i^{\tilde{T}}(X) \to C_i^T(X)$$

compatible with the boundary maps such that if $\tilde{\sigma} \in \tilde{N}^{(i)}$ then

$$\tilde{\sigma} \longmapsto \sum_{\sigma \in N^{(i)}, T(\sigma) \subseteq \tilde{T}(\tilde{\sigma})} \pm \sigma$$

where the sign depends on whether the orientations of σ and $\tilde{\sigma}$ are compatible.

Definition 2.1.7. The space $C_i(X)$ of all **piecewise linear i-chains** is the colimit of the spaces $C_i^T(X)$ under refinement. That is, a piecewise linear i-chain on X is represented by an element of $C_i^T(X)$ for some triangulation T of X, and two such elements

$$c \in C_i^T(X) \quad \text{and} \quad \tilde{c} \in C_i^{\tilde{T}}(X)$$

represent the same piecewise linear i-chain if and only if there exists a common refinement $\bar{T}$ of T and $\tilde{T}$ such that the images of c and $\tilde{c}$ in $C_i^{\bar{T}}(X)$ coincide.

The boundary maps $\partial\colon C_i^T(X) \to C_{i-1}^T(X)$ induce boundary maps

$$\partial\colon C_i(X) \to C_{i-1}(X)$$

such that $\partial^2 = 0$.

Definition 2.1.8. The **simplicial homology** of a triangulable space X is defined by

$$H_i^{\mathrm{simp}}(X) = \frac{\ker \partial\colon C_i(X) \to C_{i-1}(X)}{\mathrm{im}\ \partial\colon C_{i+1}(X) \to C_i(X)}.$$

This definition is independent of the choice of triangulation but *a priori* impossible to compute.

2.2   Singular homology

Singular homology is the most common first definition of homology. A **singular i-simplex** in a topological space X is a continuous map

$$\Sigma\colon \Delta_i \to X$$

where Δ_i is the standard i-simplex in $\mathbb{R}^i$; that is, Δ_i is the convex hull of the set of points

$$\{(0,\ldots,0),(1,0,\ldots,0),(0,1,0,\ldots,0),\ldots,(0,\ldots,0,1)\}$$

in $\mathbb{R}^i$. The space $S_i(X)$ of **singular i-chains** in X is the vector space over $\mathbb{F}$ with the set of singular i-simplices in X as basis. A singular $(i-1)$-simplex is a face of a singular i-simplex Σ if it is the composition of Σ with one of the $i+1$ maps

$$\phi_j\colon \Delta_{i-1} \to \Delta_i, \qquad 0 \le j \le i,$$

which identify Δ_{i-1} with faces of Δ_i. We define

$$\partial\Sigma = \sum_{T \text{ face of } \Sigma} \pm T$$

where the sign depends on orientations (cf. Section 2.1). If the sign is chosen correctly we get $\partial^2 = 0$ and we define the ith **singular homology group** of X with coefficients in $\mathbb{F}$ to be the quotient

$$H_i^{\text{sing}}(X) = \frac{\ker \partial \colon S_i(X) \to S_{i-1}(X)}{\text{im } \partial \colon S_{i+1}(X) \to S_i(X)}. \tag{2.1}$$

If $f \colon X \to Y$ is a continuous map between topological spaces then f induces linear maps

$$f_* \colon S_i(X) \to S_i(Y) : \sigma \mapsto f \circ \sigma$$

for any singular i-simplex Σ in X. The maps f_* are compatible with the boundary maps, and hence induce

$$f_* \colon H_*^{\text{sing}}(X) \to H_*^{\text{sing}}(Y). \tag{2.2}$$

It is not quite so obvious that if X and Y are triangulable then a continuous map $f \colon X \to Y$ induces in a natural way a linear map $f_* \colon H_*^{\text{simp}}(X) \to H_*^{\text{simp}}(Y)$ on simplicial homology. This follows however from the following important fact.

Theorem 2.2.1. *If $T \colon |N| \to X$ is any triangulation of a topological space X then there are natural isomorphisms*

$$H_*^{\text{sing}}(X) \cong H_*^{\text{simp}}(X) \cong H_*^T(X).$$

The proof is based on the simplicial approximation theorem (Spanier [163, Ch. 3 §4, Ch. 4 §6 Thm. 8]), which tells us that any singular p-simplex

$$\Sigma \colon \Delta_p \to X$$

can be 'approximated' by $\tilde{\Sigma} \colon \Delta_p \to X$ where $\tilde{\Sigma}$ is piecewise-linear with respect to the given triangulation T on X and a refinement of the obvious triangulation on Δ_p. The approximation is such that

$$\Sigma = \tilde{\Sigma} + \partial\bar{\Sigma}$$

for some $(p+1)$-chain $\bar{\Sigma}$ in $S_{p+1}(X)$. Since $\partial^2 = 0$ we get

$$\partial\Sigma = \partial\tilde{\Sigma}.$$

It follows that the natural map

$$H_*^T(X) \to H_*^{\text{sing}}(X)$$

is an isomorphism. Taking colimits, since $H_*^{\text{sing}}(X)$ is independent of T we find that

$$H_*^{\text{simp}}(X) \cong H_*^{\text{sing}}(X).$$

To define **cohomology** groups instead of homology groups we can use the dual $\partial^\vee$ of the boundary operator. Thus the ith singular cohomology group of X is

$$H^i_{\text{sing}}(X) = \frac{\ker \partial^\vee : C_i(X)^\vee \to C_{i+1}(X)^\vee}{\text{im } \partial^\vee : C_{i-1}(X)^\vee \to C_i(X)^\vee}$$

and the ith simplicial cohomology group $H^i_{\text{simp}}(Y)$ is defined similarly.

Because we are working with coefficients in a *field* $\mathbb{F}$, not an arbitrary ring, we have natural isomorphisms

$$H^i_{\text{sing}}(X) \cong (H^{\text{sing}}_i(X))^\vee$$

and

$$H^i_{\text{simp}}(X) \cong (H^{\text{simp}}_i(X))^\vee$$

between the cohomology groups and the duals of the corresponding homology groups.

2.3   Homology with closed support

We have defined the simplicial and singular homology groups of a triangulable space X using chains which are *finite* linear combinations of simplices. It is also possible to work with chains which are (formal) infinite linear combinations of simplices (Borel and Moore [22]). We get new homology groups (sometimes called Borel–Moore homology groups although we will use the term homology groups with closed support). When X is compact the two sorts of homology are canonically isomorphic.

Let $T \colon |N| \to X$ be a triangulation of X. The space

$$C^T_i((X))$$

of **locally finite i-chains** of X with respect to T is the vector space consisting of all formal linear combinations

$$\xi = \sum_{\sigma \in N^{(i)}} \xi_\sigma \sigma$$

where the coefficients ξ_σ are in the field $\mathbb{F}$. We do *not* impose the condition that only finitely many of the ξ_σ are non-zero. $C^T_i(X)$ is the subspace of $C^T_i((X))$ spanned by $N^{(i)}$ and we can identify $C^T_i((X))$ with the dual of $C^T_i(X)$ using the basis $N^{(i)}$.

We define the space $C_i((X))$ of locally finite piecewise linear i-chains on X as the colimit of the spaces $C^T_i((X))$ under refinement.

The support

$$|\xi| = \bigcup_{\xi_\sigma \neq 0} T(\sigma)$$

of a locally finite i-chain

$$\xi = \sum_{\sigma \in N^{(i)}} \xi_\sigma \sigma \in C_i^T((X))$$

is always a closed subset of X (since any simplicial complex N is locally finite). It is easy to see that the support $|\xi|$ is compact if and only if $\xi \in C_i^T(X)$. Thus i-chains $\xi \in C_i(X)$ are sometimes called i-**chains with compact support** (and the groups $H_i^{\text{simp}}(X)$ are called homology with compact support) whereas chains $\xi \in C_i^T((X))$ are called i-**chains with closed support**.

The boundary map $\partial \colon C_i^T(X) \to C_{i-1}^T(X)$ extends in the obvious way to a boundary map

$$\partial \colon C_i^T((X)) \to C_{i-1}^T((X))$$

such that $\partial^2 = 0$. There is an induced boundary map

$$\partial \colon C_i((X)) \to C_{i-1}((X)).$$

Definition 2.3.1. The **homology groups with closed support** (or Borel–Moore homology groups) $H_i^{cl}(X)$ of X are defined to be the quotients

$$H_i^{cl}(X) = \frac{\ker \partial \colon C_i((X)) \to C_{i-1}((X))}{\operatorname{im}\ \partial \colon C_{i+1}((X)) \to C_i((X))}.$$

Of course when X is compact then if $T \colon |N| \to X$ is a triangulation the simplicial complex N is finite, so

$$C_i^T((X)) = C_i^T(X).$$

Thus

$$H_i^{cl}(X) = H_i^{\text{simp}}(X) \tag{2.3}$$

when X is compact.

We can also define singular homology groups with closed support by considering locally finite singular chains: a formal linear combination

$$\xi = \sum_\sigma \xi_\sigma \sigma$$

of singular i-simplices in X is a **locally finite** singular i-chain in $S_i((X))$ if for each $x \in X$ there is an open neighbourhood U_x of x in X such that the set

$$\{\xi_\sigma | \xi_\sigma \neq 0, \sigma^{-1}(U_x) \neq \emptyset\}$$

is finite.

2.4   Conclusion

We have two different definitions of the cohomology of a (triangulable) topological space X

$$
\begin{aligned}
&H^*_{\mathrm{simp}}(X) &&\text{simplicial cohomology}\\
&H^*_{\mathrm{sing}}(X) &&\text{singular cohomology}
\end{aligned}
$$

and these are canonically isomorphic. We shall denote them both simply by $H^*(X)$. If X is a smooth manifold and we take $\mathbb{F} = \mathbb{C}$ then $H^*(X)$ is also canonically isomorphic to the de Rham cohomology $H^*_{\mathrm{dR}}(X)$ of X defined in the previous chapter.

To complete this chapter it is necessary to mention a few important properties of the cohomology $H^*(X)$ of X. First of all $H^*(X)$ has a natural *ring structure* defined by the **cup product**

$$H^i(X) \otimes H^j(X) \to H^{i+j}(X). \tag{2.4}$$

The cup product is easiest to describe when X is non-singular and we take $\mathbb{F} = \mathbb{C}$ so that $H^*(X)$ is identified with the de Rham cohomology $H^*_{\mathrm{dR}}(X)$. Then an element of $H^i(X)$ is represented by a closed i-form α on X (i.e. an i-form α satisfying $d\alpha = 0$). Similarly an element of $H^j(X)$ is represented by a closed j-form β on X. The cup product of these elements of $H^i(X)$ and $H^j(X)$ is the element of $H^{i+j}(X)$ represented by the $(i+j)$-form $\alpha \wedge \beta$. It is easy to check that this is well-defined by using the formula

$$d(\alpha \wedge \beta) = d\alpha \wedge \beta + (-1)^i \alpha \wedge d\beta.$$

Alternatively we can define the cup product using singular cohomology (see e.g. Spanier [163, Ch. 5 §6]). This definition makes the singular cohomology of *any* topological space into a ring.

The existence of a natural ring structure is one of the properties of ordinary cohomology which does *not* carry over to intersection cohomology. Another such property is the homotopy invariance of ordinary cohomology: if $f\colon X \to Y$ is a homotopy equivalence between topological spaces then the induced map

$$f^*\colon H^*_{\mathrm{sing}}(Y) \to H^*_{\mathrm{sing}}(X) \tag{2.5}$$

is an isomorphism (see e.g. Spanier [163, Thm. 4.4.9]). We shall see that this is not true in general for intersection cohomology, but that intersection cohomology is homeomorphism invariant (i.e. if f is a homeomorphism then it induces an isomorphism f^* on intersection cohomology).

A property of cohomology which carries over (though only in special circumstances) to intersection cohomology is the existence of relative cohomology. Again this can be defined in different ways corresponding to the different definitions of cohomology: let us take singular cohomology. Suppose X is a topological space and Y is a subset of X. Then the space $S_i(Y)$ of singular

i-chains in Y is a subspace of the space $S_i(X)$ of all singular i-chains in X, so we can define

$$S_i(X, Y) = \frac{S_i(X)}{S_i(Y)}.$$

Then the boundary map $\partial \colon S_i(X) \to S_{i-1}(X)$ induces a boundary map

$$\partial \colon S_i(X, Y) \to S_{i-1}(X, Y).$$

We define the ith **relative (singular) homology group** of the pair (X, Y) to be

$$H_i^{\mathrm{sing}}(X, Y) = \frac{\ker \partial \colon S_i(X, Y) \to S_{i-1}(X, Y)}{\operatorname{im} \partial \colon S_{i+1}(X, Y) \to S_i(X, Y)}.$$

This group fits into a long exact sequence of Abelian groups

$$\cdots \to H_i^{\mathrm{sing}}(Y) \to H_i^{\mathrm{sing}}(X) \to H_i^{\mathrm{sing}}(X, Y) \to H_{i-1}^{\mathrm{sing}}(Y) \to \cdots \qquad (2.6)$$

(Spanier [163, Ch. 4 §5]). Similarly we can define the ith relative (singular) cohomology groups $H^i_{\mathrm{sing}}(X, Y)$ and these fit into a long exact sequence

$$\cdots \to H^{i-1}_{\mathrm{sing}}(Y) \to H^i_{\mathrm{sing}}(X, Y) \to H^i_{\mathrm{sing}}(X) \to H^i_{\mathrm{sing}}(Y) \to \cdots \qquad (2.7)$$

2.5 Further reading

The material in this chapter is standard and there are many excellent accounts. For more details see e.g. Dold [57], Davis and Kirk [51], Greenberg [75], Hatcher [80] and Spanier [163]. The relation between singular and de Rham cohomology is nicely explained in Bott and Tu [26].

Chapter 3

Review of sheaf cohomology and derived categories

3.1 Sheaves

We have now considered simplicial and singular homology and cohomology and also de Rham cohomology for compact manifolds. In order to define two more important forms of cohomology we need to review some sheaf theory. The reader who wants to get a quick feel for intersection homology can skip this chapter and go straight on to the next, as sheaf theory is only used for the more advanced material in Chapters 7, 8 and later.

For algebraic simplicity we work with sheaves of vector spaces over a fixed field $\mathbb{F}$. However much of the material in this chapter is valid *mutatis mutandis* for sheaves of modules over a commutative ring, in particular for $\mathbb{Z}$-modules, i.e. Abelian groups.

Definition 3.1.1. A **presheaf** $\mathcal{F}$ of vector spaces over $\mathbb{F}$ on a topological space X is given by the following data:

(a) for every open subset U of X a vector space $\mathcal{F}(U)$,

(b) for every inclusion $U \subseteq V$ of open subsets of X a homomorphism

$$\rho_{VU} \colon \mathcal{F}(V) \to \mathcal{F}(U)$$

   called the restriction homomorphism, satisfying

(i) $\mathcal{F}(\emptyset) = 0$;

(ii) $\rho_{UU} \colon \mathcal{F}(U) \to \mathcal{F}(U)$ is the identity;

(iii) if $U \subseteq V \subseteq W$ then $\rho_{WU} = \rho_{VU} \circ \rho_{WV}$.

If $U \subseteq V$ are open subsets of X and $s \in \mathcal{F}(V)$ then we write $s|_U$ for

$$\rho_{VU}(s) \in \mathcal{F}(U).$$

A presheaf $\mathcal{F}$ on X is a **sheaf** if in addition it has the following property:

(iv) Let $\{V_i | i \in I\}$ be a collection of open subsets of X. Suppose that we are given elements $s_i \in \mathcal{F}(V_i)$ for all $i \in I$ satisfying

$$s_i\big|_{V_i \cap V_j} = s_j\big|_{V_i \cap V_j}$$

for all $i, j \in I$. Then there exists a *unique* $s \in \mathcal{F}(\bigcup_{i \in I} V_i)$ such that

$$s\big|_{V_i} = s_i$$

for all $i \in I$.

Examples 3.1.2. 1. A sheaf on the one point space is simply a vector space over $\mathbb{F}$.

2. Let L be any vector space over $\mathbb{F}$. The **constant sheaf** L_X on X determined by L is defined by

$$L_X(\emptyset) = \{0\}, \quad L_X(U) = \{\text{ continuous maps } f \colon U \to V\} \text{ if } U \neq \emptyset$$

with the obvious restriction homomorphisms. Here L is supposed to have the *discrete* topology. (Even if L has a natural topology such as when $L = \mathbb{C}$, we take the discrete topology when defining L_X.) Thus if U is non-empty and connected every continuous map $f \colon U \to V$ is constant so $L_X(U) = L$.

3. Let $\pi \colon Y \to X$ be continuous and define $\mathcal{F}$ by

$$
\begin{aligned}
\mathcal{F}(U) \;&=\; \{\text{continuous } \sigma \colon U \to Y \,|\, \pi \circ \sigma(x) = x, \forall x \in U\} \\
&=\; \{\text{sections of } \pi \text{ over } U\}
\end{aligned}
$$

with the obvious restriction maps. $\mathcal{F}$ is called the **sheaf of sections** of $\pi \colon Y \to X$. In general this will define a sheaf of sets but if, for example, Y is a vector bundle over X then it will be a sheaf of vector spaces.

Definition 3.1.3. Let $\mathcal{F}$ be a presheaf over X. The **stalk** $\mathcal{F}_x$ of $\mathcal{F}$ at $x \in X$ is the colimit of vector spaces

$$\mathcal{F}_x = \operatorname{colim} \{\mathcal{F}(U) | x \in U, \; U \text{ open in } X\}.$$

Thus an element of $\mathcal{F}_x$ is represented by a pair (U, s) where U is an open subset of X such that $x \in U$ and $s \in \mathcal{F}(U)$. Two pairs (U, s) and (V, t) represent the same element of $\mathcal{F}_x$ if there exists an open neighbourhood W of x in X such that $W \subseteq U \cap V$ and $s|_W = t|_W$. We write s_x for the element of $\mathcal{F}_x$ represented by (U, s).

Elements of $\mathcal{F}(U)$ are called **sections** of $\mathcal{F}$ over U. Elements of the stalk $\mathcal{F}_x$ are called **germs** of sections of $\mathcal{F}$ at x.

Definition 3.1.4. Suppose $\mathcal{F}$ and $\mathcal{G}$ are (pre)sheaves over X. A **map of (pre)sheaves** $\phi \colon \mathcal{F} \to \mathcal{G}$ is given by homomorphisms

$$\phi(U) \colon \mathcal{F}(U) \to \mathcal{G}(U)$$

for all open subsets $U \subseteq X$ such that if $V \subseteq U$ then the diagram

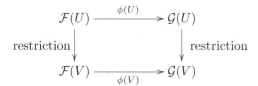

commutes. There is then an induced homomorphism $\phi_x \colon \mathcal{F}_x \to \mathcal{G}_x$ for all $x \in X$.

An important principle is that sheaves, as opposed to presheaves, are determined by their stalks. A more precise formulation is

Theorem 3.1.5. *A map $\phi \colon \mathcal{F} \to \mathcal{G}$ of presheaves is called an isomorphism if $\phi(U) \colon \mathcal{F}(U) \to \mathcal{G}(U)$ is an isomorphism for all open subsets U of X. If $\mathcal{F}$ and $\mathcal{G}$ are sheaves this is the case if and only if $\phi_x \colon \mathcal{F}_x \to \mathcal{G}_x$ is an isomorphism for all $x \in X$.*

Every sheaf is obviously a presheaf, but not conversely. However, there is a sheaf naturally associated to any presheaf.

Definition 3.1.6. Let $\mathcal{F}$ be a presheaf over X. The **sheaf $\mathcal{F}^+$ associated to $\mathcal{F}$** is defined as follows. If U is an open subset of X then $\mathcal{F}^+(U)$ is the set of functions $f \colon U \to \coprod_{x \in X} \mathcal{F}_x$ satisfying

(i) $f(x) \in \mathcal{F}_x$ for all $x \in U$ and

(ii) if $x \in U$ then there is an open neighbourhood W of x in U and there is some $s \in \mathcal{F}(W)$ such that $f(y) = s_y$ for all $y \in W$.

$\mathcal{F}^+(U)$ becomes a vector space under pointwise addition and scalar multiplication. The obvious restriction homomorphisms make $\mathcal{F}^+$ into a sheaf.

Alternatively if we put an appropriate topology on the disjoint union of stalks

$$Y = \coprod_{x \in X} \mathcal{F}_x$$

then we can define $\mathcal{F}^+$ as the sheaf of sections of $\pi \colon Y \to X$, where $\pi(s_x) = x$ if $s_x \in \mathcal{F}_x$.

Example 3.1.7. The constant sheaf V_X introduced in Example 3.1.2 is the sheaf associated to the constant presheaf whose sections over every open U are the vector space V, and all of whose restriction maps are the identity.

There is a natural map of presheaves $\phi \colon \mathcal{F} \to \mathcal{F}^+$ such that if U is open in X and $s \in \mathcal{F}(U)$ then

$$\phi(U) \colon U \to \coprod_{x \in U} \mathcal{F}_x$$

sends $x \in U$ to $s_x \in \mathcal{F}_x$. This map ϕ induces isomorphisms on stalks for all $x \in X$ and is hence an isomorphism if and only if $\mathcal{F}$ is a sheaf. It has the universal property that any map of presheaves $\psi \colon \mathcal{F} \to \mathcal{G}$ from $\mathcal{F}$ to a *sheaf* $\mathcal{G}$ over X factors uniquely as the composition of $\phi \colon \mathcal{F} \to \mathcal{F}^+$ and a map of sheaves $\theta \colon \mathcal{F}^+ \to \mathcal{G}$.

Whenever $\phi \colon \mathcal{F} \to \mathcal{G}$ is a map of sheaves over X we can define a kernel, an image and a cokernel sheaf. The definitions are as follows. The **kernel** $\ker \phi$ is the sheaf defined by

$$\ker \phi(U) = \ker\{\phi(U) \colon \mathcal{F}(U) \to \mathcal{G}(U)\}$$

with the restriction maps induced by those of $\mathcal{F}$. However the presheaf whose space of sections over U is

$$\operatorname{im}\{\phi(U) \colon \mathcal{F}(U) \to \mathcal{G}(U)\}$$

is *not* necessarily a sheaf. Since we are interested in sheaves rather than presheaves, we define the **image** $\operatorname{im}\phi$ of ϕ to be the sheaf associated to this presheaf.

A **subsheaf** $\mathcal{F}$ of $\mathcal{G}$ is a sheaf over X such that $\mathcal{F}(U)$ is a subgroup of $\mathcal{G}(U)$ for all open subsets U of X and the restriction maps of $\mathcal{F}$ are induced by those of $\mathcal{G}$. If $\mathcal{F}$ is a subsheaf of $\mathcal{G}$ then the presheaf

$$U \to \frac{\mathcal{G}(U)}{\mathcal{F}(U)}$$

is not necessarily a sheaf. We define the quotient sheaf $\mathcal{G}/\mathcal{F}$ to be the sheaf associated to this presheaf.

The kernel of ϕ is clearly a subsheaf of $\mathcal{F}$. The quotient sheaf is (isomorphic to) the image of ϕ. The universal property of a sheaf associated to a presheaf ensures that the image $\operatorname{im}\phi$ is a subsheaf of $\mathcal{G}$. The **cokernel** $\operatorname{coker}\phi$ is defined to be the quotient $\mathcal{G}/\operatorname{im}\phi$.

If this seems confusing at first then the situation on stalks is much simpler. There are natural isomorphisms

$$(\ker\phi)_x \cong \ker\phi_x, \quad (\operatorname{im}\phi)_x \cong \operatorname{im}\phi_x \text{ and } (\operatorname{coker}\phi)_x \cong \operatorname{coker}\phi_x = \mathcal{G}_x/\operatorname{im}\phi_x.$$

Definition 3.1.8. A map $\phi\colon \mathcal{F} \to \mathcal{G}$ of sheaves is **injective** if $\ker\phi = 0$, equivalently if ϕ_x injective for all $x \in X$, and **surjective** if $\operatorname{coker}\phi = 0$, equivalently if ϕ_x surjective for all $x \in X$. A diagram

$$\mathcal{E} \xrightarrow{\phi} \mathcal{F} \xrightarrow{\psi} \mathcal{G}$$

of maps of sheaves is **exact at** $\mathcal{F}$ if $\operatorname{im}\phi = \ker\psi$, equivalently if $\operatorname{im}\phi_x = \ker\psi_x$ for all $x \in X$. A **short exact sequence** is a sequence

$$0 \to \mathcal{E} \xrightarrow{\phi} \mathcal{F} \xrightarrow{\psi} \mathcal{G} \to 0,$$

which is exact at each non-zero term, in other words we have $\ker\phi = 0$, $\operatorname{im}\phi = \ker\psi$ and $\operatorname{im}\psi = \mathcal{G}$.

Example 3.1.9. Let $X = [0,1]$ and $A = [\frac{1}{3}, \frac{2}{3}]$. Put $U = X - A$. Let $\mathcal{F} = \mathbb{F}_X$ be the constant sheaf and $\mathcal{G}$ be the sheaf on X with sections $\mathcal{G}(V) = \mathcal{F}(U \cap V)$. Restriction gives a map of sheaves $\phi : \mathcal{F} \to \mathcal{G}$. If V is a connected open interval then

$$(\ker\phi)(V) = \{s \in \mathcal{F}(V) \mid s|_{U\cap V} = 0\} \cong \begin{cases} \mathbb{F} & U \cap V = \emptyset \\ 0 & U \cap V \neq \emptyset. \end{cases}$$

Note that $\phi(X)$ is not surjective since $\mathcal{F}(X) = \mathbb{F}$ but $\mathcal{G}(X) = \mathbb{F}^2$. However, ϕ induces surjections on stalks and so $\operatorname{im}\phi = \mathcal{G}$. It follows that the presheaf $V \mapsto \operatorname{im}(\phi(V))$ is not a sheaf. Exercise: check explicitly that the sheaf property 3.1.1 (iv) is not satisfied.

3.2   Čech cohomology of sheaves

Suppose we have an injection $\mathcal{F} \to \mathcal{G}$ of sheaves on a topological space X. Then we have a short exact sequence

$$0 \to \mathcal{F} \to \mathcal{G} \to \mathcal{G}/\mathcal{F} \to 0$$

of sheaves. It follows that we have an exact sequence

$$0 \to \mathcal{F}(X) \to \mathcal{G}(X) \to (\mathcal{G}/\mathcal{F})(X)$$

of global sections, but the last map need not be surjective e.g. Example 3.1.9. When is a section $s \in (\mathcal{G}/\mathcal{F})(X)$ in the image of $\mathcal{G}(X)$? The following exercise is instructive.

Exercise 3.2.1. Suppose X is compact and $\mathcal{F}$ and $\mathcal{G}$ are as above. Let $s \in (\mathcal{G}/\mathcal{F})(X)$. Show that we can find a finite cover $\{U_i\}$ of X and $t_i \in \mathcal{G}(U_i)$ such that

$$\mathcal{G}(U_i) \longrightarrow (\mathcal{G}/\mathcal{F})(U_i) : t_i \longmapsto s|_{U_i}.$$

Show that there is a unique $r_{ij} \in \mathcal{F}(U_{ij})$ with $r_{ij} \mapsto t_i - t_j \in \mathcal{G}(U_{ij})$ and that

$$r_{ij} + r_{jk} + r_{ki} = 0 \in \mathcal{F}(U_{ijk}) \tag{3.1}$$

where, for simplicity, we write U_{ij} for $U_i \cap U_j$ and t_i for $t_i|_{U_{ij}}$ etc. Finally, show that if there exist $r_i \in \mathcal{F}(U_i)$ with

$$r_{ij} = r_i - r_j \in \mathcal{F}(U_{ij}) \tag{3.2}$$

then there exists a lift $t \in \mathcal{G}(X)$ of s, i.e. t maps to s under $\mathcal{G}(X) \to (\mathcal{G}/\mathcal{F})(X)$.

There is a potential obstruction to lifting a global section of $\mathcal{G}/\mathcal{F}$ to a global section of $\mathcal{G}$ consisting of collections $\{r_{ij} \in \mathcal{F}(U_{ij})\}$ satisfying (3.1) but not arising from a collection $\{r_i \in \mathcal{F}(U_i)\}$ as in (3.2). Thus it is reasonable to expect that we have an exact sequence

$$0 \to \mathcal{F}(X) \to \mathcal{G}(X) \to (\mathcal{G}/\mathcal{F})(X) \to \check{H}^1(X; \mathcal{F})$$

where $\check{H}^1(X; \mathcal{F})$ is a 'space of obstructions' to lifting global sections of $\mathcal{G}/\mathcal{F}$ to global sections of $\mathcal{G}$ *which depends only on $\mathcal{F}$*. Here is the definition.

Let $\mathcal{F}$ be a sheaf of vector spaces over $\mathbb{F}$ on a topological space X and let $\mathcal{U} = \{U_i | i \in I\}$ be an open covering of X. For each $p \geq 0$ let $I^{(p)}$ be the set of all subsets of I with precisely $p + 1$ elements. If

$$K = \{i_0, ..., i_p\} \in I^{(p)}$$

then put $U_K = U_{i_0} \cap U_{i_1} \cap \ldots \cap U_{I_p}$. Let

$$C^p(\mathcal{U}; \mathcal{F}) = \prod_{K \in I^{(p)}} \mathcal{F}(U_K).$$

Then $C^p(\mathcal{U}; \mathcal{F})$ is a vector space over $\mathbb{F}$. An element $\alpha \in C^p(\mathcal{U}; \mathcal{F})$ is determined by giving elements $\alpha_K \in \mathcal{F}(U_K)$ for each $K \in I^{(p)}$.

For each $K \in I^{(p)}$ choose an orientation of K, i.e. an ordering of the elements of K up to even permutations. Define a coboundary map

$$d \colon C^p(\mathcal{U}; \mathcal{F}) \to C^{p+1}(\mathcal{U}; \mathcal{F})$$

as follows. If $K = \{i_0, \ldots, i_{p+1}\} \in I^{(p+1)}$ set

$$(d\alpha)_K = \sum_{j=0}^{p+1} \pm \alpha_{K-\{i_j\}}|_{U_K}$$

where the sign $\pm$ depends on whether or not the orientation chosen for K coincides with the orientation chosen for $K - \{i_j\}$ with i_j placed at the beginning. It is easy to check that

$$d^2 = 0$$

so we can define

$$\check{H}^p(\mathcal{U}; \mathcal{F}) = \frac{\ker d \colon C^p(\mathcal{U}; \mathcal{F}) \to C^{p+1}(\mathcal{U}; \mathcal{F})}{\operatorname{im} d \colon C^{p-1}(\mathcal{U}; \mathcal{F}) \to C^p(\mathcal{U}; \mathcal{F})}. \tag{3.3}$$

An open covering $\mathcal{V}$ of X is called a **refinement** of $\mathcal{U}$ if for every $V \in \mathcal{V}$ there exists some $U_V \in \mathcal{U}$ such that $V \subseteq U_V$. Then there is a map

$$C^p(\mathcal{U}; \mathcal{F}) \to C^p(\mathcal{V}; \mathcal{F})$$

induced by the restriction maps of $\mathcal{F}$ which commutes with coboundary maps. We define

$$C^p(X; \mathcal{F})$$

to be the colimit of $C^p(\mathcal{U}; \mathcal{F})$ with respect to refinement. That is, every element of $C^p(X; \mathcal{F})$ is represented by an element of $C^p(\mathcal{U}; \mathcal{F})$ for some open covering $\mathcal{U}$, and elements of $C^p(\mathcal{U}; \mathcal{F})$ and $C^p(\mathcal{V}; \mathcal{F})$ represent the same element of $C^p(X; \mathcal{F})$ if they map to the same element of $C^p(\mathcal{W}; \mathcal{F})$ for some common refinement $\mathcal{W}$ of $\mathcal{U}$ and $\mathcal{V}$.

The coboundary maps

$$d \colon C^p(\mathcal{U}; \mathcal{F}) \to C^{p+1}(\mathcal{U}; \mathcal{F})$$

induce coboundary maps

$$d \colon C^p(X; \mathcal{F}) \to C^{p+1}(X; \mathcal{F}).$$

The pth **Čech cohomology** group of X with coefficients in $\mathcal{F}$ is, by definition, the quotient

$$\check{H}^p(X; \mathcal{F}) = \frac{\ker d \colon C^p(X; \mathcal{F}) \to C^{p+1}(X; \mathcal{F})}{\operatorname{im} d \colon C^{p-1}(X; \mathcal{F}) \to C^p(X; \mathcal{F})}.$$

It is a vector space over the field $\mathbb{F}$.

Example 3.2.2. An element $\alpha \in C^0(\mathcal{U}; \mathcal{F}) = \prod_{i \in I} \mathcal{F}(U_i)$ satisfies $d\alpha = 0$ if and only if, for each i and j

$$\alpha_i|_{U_i \cap U_j} = \alpha_j|_{U_i \cap U_j},$$

i.e. if and only if the α_i patch together to form a global section. Hence

$$\check{H}^0(X; \mathcal{F}) \cong \mathcal{F}(X).$$

The definition of Čech cohomology is functorial; if we have a map $\mathcal{F} \to \mathcal{G}$ of sheaves then there are induced maps $C^p(X; \mathcal{F}) \to C^p(X; \mathcal{G})$ which commute with the coboundaries and thence induce maps on Čech cohomology groups. In particular if

$$0 \to \mathcal{F} \to \mathcal{G} \to \mathcal{G}/\mathcal{F} \to 0$$

is a short exact sequence of sheaves then we obtain short exact sequences

$$0 \to C^p(X; \mathcal{F}) \to C^p(X; \mathcal{G}) \to C^p(X; \mathcal{G}/\mathcal{F}) \to 0$$

of vector spaces for each $p \geq 0$. A standard result in homological algebra (see e.g. Spanier [163, Ch. 4 §5]) then tells us that we obtain an exact sequence of Čech cohomology groups

$$0 \to \check{H}^0(X; \mathcal{F}) \to \check{H}^0(X; \mathcal{G}) \to \check{H}^0(X; \mathcal{G}/\mathcal{F}) \to \check{H}^1(X; \mathcal{F}) \to \cdots$$

Combined with Example 3.2.2 this yields the promised interpretation of Čech cohomology groups as spaces of obstructions to lifting sections.

If X is triangulable we can always choose an open cover $\mathcal{U}$ so that

$$\check{H}^p(\mathcal{U}; \mathcal{F}) = \check{H}^p(X; \mathcal{F}) \,.$$

This follows from the proof of the following proposition.

Proposition 3.2.3. *If X is triangulable then $H^*_{\mathrm{simp}}(X; \mathbb{F}) \cong \check{H}^*(X; \mathbb{F}_X)$ where $\mathbb{F}_X$ is the constant sheaf on X determined by the field $\mathbb{F}$.*

Sketch proof. Consider a triangulation $T: |N| \to X$ of X. Let V be the set of vertices of N. If $\sigma \in N$ let

$$\sigma^\circ = \sigma - \bigcup_{\tau < \sigma} \tau$$

be the interior of σ. (Here we write $\tau < \sigma$ to mean τ is a face of sigma and $\tau \neq \sigma$.) For each $v \in V$ define the **star** of v to be the open set

$$U_v = \bigcup_{\sigma \in N, v \in \sigma} T(\sigma^\circ).$$

Then $\mathcal{U} = \{U_v | v \in V\}$ is an open covering of X. Furthermore, if K is the set $\{v_0, \ldots, v_p\} \in V^{(p)}$ then

$$U_K = U_{v_0} \cap \ldots \cap U_{v_p}$$

is non-empty and connected if $v_0, \ldots, v_p$ are the vertices of a p-simplex in N, and is empty otherwise. Thus the constant sheaf $\mathbb{F}_X$ satisfies

$$\mathbb{F}_X(U_K) = \begin{cases} \mathbb{F} & \text{if } K \text{ spans a } p\text{-simplex in } N \\ 0 & \text{otherwise.} \end{cases}$$

So, given a Čech cochain $\alpha \in C^p(\mathcal{U}; \mathbb{C}_X)$, or equivalently given elements

$$\alpha_K \in \mathbb{C}_X(U_K)$$

for all $K \in I^{(p)}$, we can define a simplicial cochain

$$\phi(\alpha) \in \left(C_p^T(X)\right)^{\vee}$$

by putting

$$\phi(\alpha) \cdot \tau = \pm \alpha_{\{v_0,\ldots,v_p\}} \in \mathbb{F}_X(U_{\{v_0,\ldots,v_p\}}) = \mathbb{F}$$

if τ is the p-simplex with the vertices $v_0, \ldots, v_p$, and extending linearly. The sign depends on whether the orientation chosen for τ is the same as that chosen for $K = \{v_0, ..., v_p\}$. We thus get an isomorphism

$$\phi \colon C^p(\mathcal{U}; \mathbb{F}_X) \to \left(C_p^T(X)\right)^{\vee}$$

which respects the coboundary maps and hence induces an isomorphism

$$\check{H}^*(\mathcal{U}; \mathbb{F}_X) \to H_T^*(X).$$

Since we can refine T to make $\mathcal{U}$ arbitrarily fine we get, in the limit, an isomorphism

$$\check{H}^*(X; \mathbb{F}_X) \cong H_{\text{simp}}^*(X; \mathbb{F}).$$

$\square$

3.3 Hypercohomology

Later we will need to associate cohomology groups not just to a single sheaf but also to a **complex $\mathcal{F}^\bullet$ of sheaves** — that is a sequence

$$\cdots \to \mathcal{F}^i \xrightarrow{d} \mathcal{F}^{i+1} \to \cdots$$

of sheaves $\mathcal{F}^i$ indexed by $i \in \mathbb{Z}$ and sheaf maps $\mathcal{F}^i \to \mathcal{F}^{i+1}$ (all of which we denote by d) in which the composite of any two consecutive maps is 0, or, more succinctly, $d^2 = 0$. In order to distinguish the groups associated to a complex we will refer to them as *hyper*cohomology groups. They are defined as follows.

Let $\mathcal{F}^\bullet$ be a complex of sheaves of vector spaces over $\mathbb{F}$ on a topological space X and let $\mathcal{U}$ be an open cover of X. Let

$$C^p(\mathcal{U}; \mathcal{F}^q)$$

be the space of Čech p-cochains over $\mathcal{U}$ with coefficients in $\mathcal{F}^q$. We have a boundary map

$$\delta_1 \colon C^p(\mathcal{U}; \mathcal{F}^q) \to C^{p+1}(\mathcal{U}; \mathcal{F}^q)$$

and the sheaf complex differential $d\colon \mathcal{F}^q \to \mathcal{F}^{q+1}$ induces

$$\delta_2\colon C^p(\mathcal{U},\mathcal{F}^q) \to C^p(\mathcal{U},\mathcal{F}^{q+1})$$

satisfying $\delta_1^2 = 0 = \delta_2^2$ and $\delta_1\delta_2 = \delta_2\delta_1$.

Passing to the colimit with respect to refinement of open covers of X we obtain a vector space

$$C^{p,q} = \operatorname{colim} C^p(\mathcal{U},\mathcal{F}^q)$$

with boundary maps $\delta_1\colon C^{p,q} \to C^{p+1,q}$ and $\delta_2\colon C^{p,q} \to C^{p,q+1}$. We define the Čech **hypercohomology** $H^*(X;\mathcal{F}^\bullet)$ of $\mathcal{F}^\bullet$ to be the cohomology of the complex $(K^\bullet, d)$ where

$$K^n = \bigoplus_{p+q=n} C^{p,q}$$

and $d = \delta_1 + (-1)^p \delta_2$ on $C^{p,q}$. That is

$$\check{H}^n(X;\mathcal{F}^\bullet) = \frac{\ker d\colon K^n \to K^{n+1}}{\operatorname{im}\, d\colon K^{n-1} \to K^n}. \tag{3.4}$$

Example 3.3.1. Any sheaf $\mathcal{E}$ can be thought of as a complex

$$\cdots \to 0 \to \mathcal{E} \to 0 \to \cdots$$

with the sheaf placed in degree zero (and nothing in the other degrees). The hypercohomology of this complex is simply the cohomology of the sheaf $\mathcal{E}$.

3.4   Functors and exactness

The sheaves on a topological space X are the objects of an **Abelian category** $\mathrm{Sh}(X)$. This means that

1. The maps $\operatorname{Hom}_{\mathrm{Sh}(X)}(\mathcal{F},\mathcal{G})$ from $\mathcal{F}$ to $\mathcal{G}$ form an Abelian group[1] under the operation

$$(\phi + \psi)(U) : \mathcal{F}(U) \;\longrightarrow\; \mathcal{G}(U)$$
$$s \;\longmapsto\; \phi(U)(s) + \psi(U)(s)$$

   and composition $\operatorname{Hom}_{\mathrm{Sh}(X)}(\mathcal{E},\mathcal{F}) \times \operatorname{Hom}_{\mathrm{Sh}(X)}(\mathcal{F},\mathcal{G}) \to \operatorname{Hom}_{\mathrm{Sh}(X)}(\mathcal{E},\mathcal{G})$ is biadditive;

2. There is a zero sheaf 0 (whose sections over each open U are the 0 vector space);

3. We can form the direct sum $\mathcal{F} \oplus \mathcal{G}$ of sheaves by setting $(\mathcal{F} \oplus \mathcal{G})(U) = \mathcal{F}(U) \oplus \mathcal{G}(U)$;

---

[1] Indeed, since we work with sheaves of vector spaces, the maps form a vector space.

4. The kernel and cokernel of a sheaf map $\phi : \mathcal{F} \to \mathcal{G}$ satisfy the universal properties represented by the diagrams

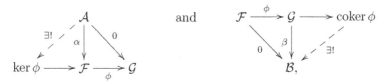

i.e. if $\phi \circ \alpha = 0$ then α factorises uniquely through $\ker \phi \to \mathcal{F}$ and if $\beta \circ \phi = 0$ then β factorises uniquely through $\mathcal{G} \to \operatorname{coker} \phi$;

5. The image of a sheaf map $\phi : \mathcal{F} \to \mathcal{G}$ fits into short exact sequences

$$0 \to \ker \phi \to \mathcal{F} \to \operatorname{im} \phi \to 0 \quad \text{and} \quad 0 \to \operatorname{im} \phi \to \mathcal{G} \to \operatorname{coker} \phi \to 0.$$

So far we have been interested in sheaves on a single space X but now we consider relationships between sheaves on different spaces.

Definition 3.4.1. A **covariant** (respectively **contravariant**) functor

$$F : \operatorname{Sh}(X) \to \operatorname{Sh}(Y)$$

from the category of sheaves on X to the category of sheaves on Y is a rule which assigns to each sheaf $\mathcal{F}$ on X a sheaf $F(\mathcal{F})$ on Y and to each map of sheaves $\phi \colon \mathcal{F} \to \mathcal{G}$ on X a map $F(\phi) \colon F(\mathcal{F}) \to F(\mathcal{G})$ (respectively $F(\phi) \colon F(\mathcal{G}) \to F(\mathcal{F})$) of sheaves on Y satisfying

i) $F(1_{\mathcal{F}}) = 1_{F(\mathcal{F})}$ where $1_{\mathcal{F}}$ and $1_{F(\mathcal{F})}$ are the identity maps on $\mathcal{F}$ and $F(\mathcal{F})$;

ii) $F(\phi \circ \psi) = F(\phi) \circ F(\psi)$ (respectively $F(\phi \circ \psi) = F(\psi) \circ F(\phi)$).

The functor F is **additive** if furthermore

iii) when ϕ, ψ are both maps of sheaves $\mathcal{F} \to \mathcal{G}$ then

$$F(\phi + \psi) = F(\phi) + F(\psi)$$

where the map of sheaves $\phi + \psi \colon \mathcal{F} \to \mathcal{G}$ is defined by

$$(\phi + \psi)(U)(s) = \phi(U)(s) + \psi(U)(s)$$

for all U open in X and $s \in F(U)$.

The functor F is **exact** if in addition

iv) given a short exact sequence $0 \to \mathcal{E} \xrightarrow{\phi} \mathcal{F} \xrightarrow{\psi} \mathcal{G} \to 0$ of maps of sheaves over X the sequence

$$0 \to F(\mathcal{E}) \xrightarrow{F(\phi)} F(\mathcal{F}) \xrightarrow{F(\psi)} F(\mathcal{G}) \to 0$$

(or $0 \to F(\mathcal{G}) \to F(\mathcal{F}) \to F(\mathcal{E}) \to 0$ in the contravariant case) is an exact sequence of Abelian groups.

In brief, an exact functor is one which preserves the Abelian category structure.

An additive functor F is called **left** (respectively **right**) **exact** if the sequence of Abelian groups obtained from any short exact sequence of sheaves via F is left (respectively right) exact: that is, we drop the condition that the second map should be surjective (respectively that the first map should be injective).

In the remainder of this section we define some important examples of functors and discuss their exactness properties.

A continuous map $f \colon X \to Y$ induces (covariant) functors

$$\mathrm{Sh}(X) \underset{f^*}{\overset{f_*}{\rightleftarrows}} \mathrm{Sh}(Y)$$

defined as follows. For a sheaf $\mathcal{F}$ on X we define the **pushforward** $f_*\mathcal{F}$ to be the sheaf on Y with sections

$$f_*\mathcal{F}(V) = \mathcal{F}\big(f^{-1}(V)\big)$$

for open subsets V of Y. For a sheaf $\mathcal{G}$ on Y we define the **pullback** $f^*\mathcal{G}$ to be the sheaf on X associated to the presheaf with sections

$$U \mapsto \mathrm{colim}\,_{V \supset f(U)}\mathcal{G}(V)$$

where V runs over all open subsets of Y containing $f(U)$. If $f \colon X \to Y$ is the inclusion of a subset X of Y in Y then $f^*\mathcal{G}$ is called the **restriction** of $\mathcal{G}$ to X. We leave the reader to define f_* and f^* on maps of sheaves.

Definition 3.4.2. If $s \in \mathcal{F}(U)$ is a section of $\mathcal{F}$ over U then the **support** $|s|$ of s is the closure in U of the subset

$$\{x \in U \,|\, s_x \neq 0\}$$

where s_x is the image of s in the stalk $\mathcal{F}_x$. If $|s|$ is compact then s is said to have compact support.

We can also define a **pushforward with proper supports** functor

$$f_! \colon \mathrm{Sh}(X) \to \mathrm{Sh}(Y)$$

by

$$f_!\mathcal{F}(V) = \{s \in \mathcal{F}(f^{-1}V) \mid f \colon |s| \to Y \text{ is a proper map}\}$$

(where a continuous map is proper if and only if the inverse image of every compact subset is compact).

Examples 3.4.3.   1. Every space has a unique map $p_X \colon X \to \mathrm{pt}$ to the point. We can identify sheaves on the point with vector spaces over

$\mathbb{F}$ and, under this identification, $p_{X*}\mathcal{F} = \mathcal{F}(X)$ is the space of global sections and $p_{X!}\mathcal{F}$ is the space of global sections with compact support. For a vector space L the pullback $p_X^* L = L_X$ is the constant sheaf with stalk L.

2. If $f : \{x\} \to X$ is the inclusion of a point and L a vector space then $f_* L \cong f_! L$ is a **skyscraper sheaf** at x (so-called because the stalks are zero except for the stalk at x).

The pullback $f^* \mathcal{F}$ of a sheaf $\mathcal{F}$ on X is the stalk $\mathcal{F}_x$ at x.

Exercise 3.4.4. 1. Show that the stalk $(f^* \mathcal{G})_x \cong \mathcal{G}_{f(x)}$ and deduce that f^* is an exact functor.

2. Show that f_* and $f_!$ are left exact (but not exact). [Hint: consider Example 3.1.9 and take f to be the map to a point.]

The pushforward and pullback of sheaves are closely related.

Proposition 3.4.5 (Iversen [84, Thm. 4.8]). *The pushforward f_* is* **right adjoint** *to the pullback f^*; in other words there is a natural isomorphism*

$$\operatorname{Hom}_{\operatorname{Sh}(X)}(f^* \mathcal{G}, \mathcal{F}) \cong \operatorname{Hom}_{\operatorname{Sh}(Y)}(\mathcal{G}, f_* \mathcal{F})$$

for any sheaf $\mathcal{F}$ on X and $\mathcal{G}$ on Y.

Another class of examples arises as follows. Fix a sheaf $\mathcal{F}$ on X and consider the map

$$\operatorname{Hom}_{\operatorname{Sh}(X)}(\mathcal{F}, -) : \operatorname{Sh}(X) \to \mathbb{F}\text{-VS} : \mathcal{G} \mapsto \operatorname{Hom}_{\operatorname{Sh}(X)}(\mathcal{F}, \mathcal{G})$$

where $\mathbb{F}$-VS is the category of vector spaces over $\mathbb{F}$. This becomes a (covariant) functor if we define $\operatorname{Hom}_{\operatorname{Sh}(X)}(\mathcal{F}, \phi)$ for a map $\phi : \mathcal{G} \to \mathcal{G}'$ to be the map of vector spaces

$$\operatorname{Hom}_{\operatorname{Sh}(X)}(\mathcal{F}, \mathcal{G}) \to \operatorname{Hom}_{\operatorname{Sh}(X)}(\mathcal{F}, \mathcal{G}') : \alpha \mapsto \phi \circ \alpha$$

arising from composition.

Similarly we can consider $\mathcal{F} \mapsto \operatorname{Hom}_{\operatorname{Sh}(X)}(\mathcal{F}, \mathcal{G})$. A sheaf map $\phi : \mathcal{F} \to \mathcal{F}'$ induces a map of vector spaces

$$\operatorname{Hom}_{\operatorname{Sh}(X)}(\mathcal{F}', \mathcal{G}) \to \operatorname{Hom}_{\operatorname{Sh}(X)}(\mathcal{F}, \mathcal{G}) : \alpha \mapsto \alpha \circ \phi$$

so this is a contravariant functor.

Exercise 3.4.6. Show that $\operatorname{Hom}_{\operatorname{Sh}(X)}(-, \mathcal{G})$ is left exact. Show that

$$\operatorname{Hom}_{\operatorname{Sh}(X)}(\mathbb{F}_X, \mathcal{G}) \cong \mathcal{G}(X)$$

and deduce that $\operatorname{Hom}_{\operatorname{Sh}(X)}(-, \mathcal{G})$ need not be exact. Deduce further that $\operatorname{Hom}_{\operatorname{Sh}(X)}(\mathcal{F}, -)$ is left exact but not necessarily exact.

Adhering to the spirit of sheaf theory we can consider all open sets at once: fix $\mathcal{F}$ and $\mathcal{G}$ and define, for open $U \subset X$,

$$\mathcal{H}om_{\text{Sh}(X)}(\mathcal{F},\mathcal{G})(U) = \text{Hom}_{\text{Sh}(X)}(\mathcal{F}|_U, \mathcal{G}|_U)$$

where $\mathcal{F}|_U$ is the restriction of $\mathcal{F}$ to U, i.e. the sheaf with sections $V \mapsto \mathcal{F}(U \cap V)$. This defines a sheaf $\mathcal{H}om_{\text{Sh}(X)}(\mathcal{F},\mathcal{G})$ on X and so we have (what turn out to be left exact) functors from $\text{Sh}(X)$ to itself

$$\mathcal{F} \mapsto \mathcal{H}om_{\text{Sh}(X)}(\mathcal{F},\mathcal{G}) \text{ and } \mathcal{G} \mapsto \mathcal{H}om_{\text{Sh}(X)}(\mathcal{F},\mathcal{G}).$$

In particular note that there is a natural isomorphism

$$\mathcal{H}om_{\text{Sh}(X)}(\mathbb{F}_X, \mathcal{G}) \cong \mathcal{G}. \tag{3.5}$$

3.5   Resolutions of sheaves and of complexes

Suppose $F : \text{Sh}(X) \to \text{Sh}(Y)$ is a left exact, but not exact, functor. Then there are short exact sequences of sheaves in $\text{Sh}(X)$ whose image in $\text{Sh}(Y)$ under F are no longer exact. It is natural to ask whether we can identify 'good' classes of short exact sequences whose images *are* exact?

Definition 3.5.1. A sheaf $\mathcal{I}$ is called **injective** if the contravariant functor $\text{Hom}_{\text{Sh}(X)}(-,\mathcal{I})$ from $\text{Sh}(X)$ to $\mathbb{F}$-VS is exact. This functor is left exact for any sheaf $\mathcal{I}$, so $\mathcal{I}$ is injective if and only if given $\psi \colon \mathcal{F} \to \mathcal{G}$ with $\ker \psi = 0$, every sheaf map $\mathcal{F} \to \mathcal{I}$ extends (not necessarily uniquely) to a map $\mathcal{G} \to \mathcal{I}$ such that the diagram

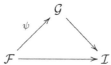

commutes.

Example 3.5.2.   1. Any vector space in $\mathbb{F}$-VS, considered as a sheaf on the point space, is injective.

  2. If $\mathcal{F}$ is any sheaf then the assignment

  $$U \mapsto \coprod_{x \in U} \mathcal{F}_x$$

  defines an injective sheaf $I^0(\mathcal{F})$.

Theorem 3.5.3. *If $\mathcal{E}$ is an injective sheaf on X and*

$$0 \to \mathcal{E} \xrightarrow{\phi} \mathcal{F} \xrightarrow{\psi} \mathcal{G} \to 0$$

is a short exact sequence then for any left exact functor $F : \text{Sh}(X) \to \text{Sh}(Y)$ the image

$$0 \to F(\mathcal{E}) \xrightarrow{F(\phi)} F(\mathcal{F}) \xrightarrow{F(\psi)} F(\mathcal{G}) \to 0$$

is also short exact.

A proof can be found in Iversen [84, I.7] or in Hartshorne [79, Ch. III, Thm. 1.1A]. The category of sheaves on a topological space 'has enough injectives': more precisely

Proposition 3.5.4 (Hartshorne [79, Ch. III, 2.3]). *If $\mathcal{F}$ is a sheaf on X then there is an exact sequence (the* **Godement resolution***)*

$$0 \to \mathcal{F} \to I^0(\mathcal{F}) \to I^1(\mathcal{F}) \to \cdots$$

of sheaves in which $I^j(\mathcal{F})$ is injective for all j. Furthermore the construction is functorial, i.e. a sheaf map $\phi : \mathcal{F} \to \mathcal{G}$ induces maps $I^j(\phi)$ between the Godement resolutions so that

$$
\begin{array}{ccccccccc}
0 & \longrightarrow & \mathcal{F} & \longrightarrow & I^0(\mathcal{F}) & \longrightarrow & I^1(\mathcal{F}) & \longrightarrow & \cdots \\
& & \downarrow{\scriptstyle\phi} & & \downarrow{\scriptstyle I^0(\phi)} & & \downarrow{\scriptstyle I^1(\phi)} & & \\
0 & \longrightarrow & \mathcal{G} & \longrightarrow & I^0(\mathcal{G}) & \longrightarrow & I^1(\mathcal{G}) & \longrightarrow & \cdots
\end{array}
$$

commutes.

Sketch proof. Define $I^0(\mathcal{F})$ as in Example 3.5. It is easy to see that $\mathcal{F}$ injects into $I^0(\mathcal{F})$ via

$$\mathcal{F}(U) \to I^0(\mathcal{F})(U) : s \mapsto \coprod_{x \in U} s_x.$$

Now perform the same procedure on the cokernel $I^0(\mathcal{F})/\mathcal{F}$ of this injection, i.e. put

$$I^1(\mathcal{F}) = I^0(\mathcal{F})/\mathcal{F}$$

and continue inductively. $\qquad\qquad\square$

The Godement resolution is an example of an injective resolution of a sheaf. However, we will need the more general notion of an injective resolution of a *complex of sheaves* and so we postpone the formal definition of injective resolution in favour of introducing some important terminology.

Definition 3.5.5. A **cochain map** $\phi^\bullet : \mathcal{F}^\bullet \to \mathcal{G}^\bullet$ between complexes of sheaves is a collection $\phi^i : \mathcal{F}^i \to \mathcal{G}^i$ of sheaf maps, indexed by $i \in \mathbb{Z}$, such that each square

$$
\begin{array}{ccc}
\mathcal{F}^i & \xrightarrow{\ d\ } & \mathcal{F}^{i+1} \\
\downarrow{\scriptstyle \phi^i} & & \downarrow{\scriptstyle \phi^{i+1}} \\
\mathcal{G}^i & \xrightarrow[\ d\ ]{} & \mathcal{G}^{i+1}
\end{array}
$$

commutes. A cochain map $\phi^\bullet : \mathcal{F}^\bullet \to \mathcal{G}^\bullet$ is a **quasi-isomorphism** if it induces isomorphisms of sheaves

$$\mathcal{H}^i(\mathcal{F}^\bullet) \to \mathcal{H}^i(\mathcal{G}^\bullet) \qquad \forall i$$

where

$$\mathcal{H}^i(\mathcal{F}^\bullet) = \frac{\ker : \mathcal{F}^i \to \mathcal{F}^{i+1}}{\mathrm{im} : \mathcal{F}^{i-1} \to \mathcal{F}^i}$$

is the ith **cohomology sheaf** of the complex $\mathcal{F}^\bullet$.

Note that the stalk of the cohomology sheaf at $x \in X$ is the cohomology of the complex of stalks, i.e. $\mathcal{H}^i(\mathcal{F}^\bullet)_x = H^i(\mathcal{F}^\bullet_x)$. Hence $\phi^\bullet$ is a quasi-isomorphism if and only if the maps

$$H^i(\mathcal{F}^\bullet_x) \to H^i(\mathcal{G}^\bullet_x)$$

are all isomorphisms.

Definition 3.5.6. An **injective resolution** of a complex $\mathcal{F}^\bullet$ of sheaves on X is a complex $\mathcal{I}^\bullet$ of injective sheaves together with a quasi-isomorphism $\mathcal{F}^\bullet \to \mathcal{I}^\bullet$.

Example 3.5.7. The Godement resolution determines a quasi-isomorphism

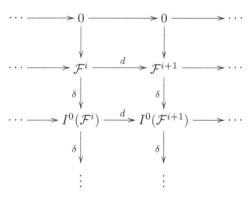

and so defines an injective resolution of the sheaf $\mathcal{F}$.

A complex of sheaves $\mathcal{F}^\bullet$ is **bounded** if there exists $N \in \mathbb{N}$ with $\mathcal{F}^n = 0$ for $|n| \geq N$.

Proposition 3.5.8. *Every bounded complex $\mathcal{F}^\bullet$ of sheaves on X has an injective resolution.*

Sketch proof. The Godement resolution provides us with injective resolutions of each $\mathcal{F}^i$ which fit into a double complex whose columns are exact sequences. A diagram chase — sometimes called the Double Complex Theorem see e.g. Bott and Tu [26, Pf. of Prop. 8.8] — shows that the compositions

$$\mathcal{F}^p \to I^0(\mathcal{F}^p) \hookrightarrow \bigoplus_{q \geq 0} I^q(\mathcal{F}^{p-q})$$

induce a quasi-isomorphism from $\mathcal{F}^\bullet$ to the **total complex** $\oplus_{q \geq 0} I^q(\mathcal{F}^{\bullet-q})$. The differential on the total complex is given on each term of the direct sum (which is finite by the boundedness assumption on $\mathcal{F}^\bullet$) by

$$d \oplus (-1)^{p+q}\delta : I^q(\mathcal{F}^{p-q}) \quad \to \quad I^q(\mathcal{F}^{(p+1)-q}) \oplus I^{(q+1)}(\mathcal{F}^{p-q})$$

$$\hookrightarrow \bigoplus_{r \geq 0} I^r(\mathcal{F}^{p+1-r}).$$

Since the direct sum of injective sheaves is injective the total complex is a complex of injective sheaves. □

Injective resolutions are not unique, but any two injective resolutions of a complex are homotopy equivalent in the following sense.

Definition 3.5.9. Cochain maps $\phi^\bullet$ and $\psi^\bullet$ from $\mathcal{F}^\bullet$ to $\mathcal{G}^\bullet$ are **homotopic** if there are maps of sheaves $\eta^i : \mathcal{F}^i \to \mathcal{G}^{i-1}$ such that

$$\phi^i - \psi^i = d \circ \eta^i + \eta^i \circ d.$$

Two complexes $\mathcal{F}^\bullet$ and $\mathcal{G}^\bullet$ are **homotopy equivalent** if there are maps $\phi^\bullet : \mathcal{F}^\bullet \to \mathcal{G}^\bullet$ and $\psi^\bullet : \mathcal{G}^\bullet \to \mathcal{F}^\bullet$ such that both $\phi^\bullet \circ \psi^\bullet$ and $\psi^\bullet \circ \phi^\bullet$ are homotopic to the identity.

The proof that any two injective resolutions are homotopy equivalent is an extended exercise in the use of the universal property of injective sheaves. Note that it easily follows that injective resolutions of quasi-isomorphic complexes are homotopy equivalent.

In a sense which will become clear, we will use an injective resolution as a surrogate for the original complex. The fact that any two injective resolutions are homotopy equivalent means that, for many purposes, it does not matter which injective resolution we use. However, for clarity and definiteness, henceforth we fix a functorial choice of injective resolution, namely that arising from the Godement resolution of a sheaf as in the proof of Proposition 3.5.8. To be explicit, for each complex of sheaves $\mathcal{F}$ on X we assign an injective resolution $\mathcal{F}^\bullet \to I^\bullet(\mathcal{F})$ and for each cochain map $\phi^\bullet : \mathcal{F}^\bullet \to \mathcal{G}^\bullet$ we assign a cochain map $I^\bullet(\phi) : I^\bullet(\mathcal{F}) \to I^\bullet(\mathcal{G})$ in such a way that $I^\bullet(\mathrm{id}) = \mathrm{id}$ and $I^\bullet(\phi \circ \psi) = I^\bullet(\phi) \circ I^\bullet(\psi)$.

3.6   Cohomology and hypercohomology via derived functors

We are now ready to give our second definition of sheaf cohomology. Recall that we can describe the global sections functor $\mathcal{F} \mapsto \mathcal{F}(X)$ as the pushforward

$$p_{X*} : \mathrm{Sh}(X) \to \mathbb{F}\text{-VS}$$

to a point. This is a left exact functor (see Exercise 3.4.4). Since p_{X*} is a functor it takes complexes of sheaves on X to complexes of vector spaces. The ith **cohomology** of a sheaf $\mathcal{F}$ on X is defined to be

$$H^i(X; \mathcal{F}) = H^i(p_{X*}I^\bullet(\mathcal{F})) = \frac{\ker : p_{X*}I^i(\mathcal{F}) \to p_{X*}I^{i+1}(\mathcal{F})}{\mathrm{im} : p_{X*}I^{i-1}(\mathcal{F}) \to p_{X*}I^i(\mathcal{F})}$$

where $\mathcal{F} \to I^\bullet(\mathcal{F})$ is the Godement resolution. Precisely the same definition serves for complexes; the ith **hypercohomology** of a complex $\mathcal{F}^\bullet$ of sheaves on X is defined by

$$H^i(X; \mathcal{F}) = H^i(p_{X*}I^\bullet(\mathcal{F}))$$

where $\mathcal{F} \to I^\bullet(\mathcal{F})$ is the chosen injective resolution. The definition of the cohomology of a sheaf is simply the special case in which the complex is zero except in one degree.

Remark 3.6.1. If we had chosen a different injective resolution $\mathcal{F}^\bullet \to \mathcal{I}^\bullet$ then $\mathcal{I}^\bullet$ and $I^i(\mathcal{F})$ would be homotopy equivalent complexes of sheaves. It follows that $p_{X*}\mathcal{I}^\bullet$ and $p_{X*}I^i(\mathcal{F})$ are homotopy equivalent complexes of vector spaces and therefore that they have isomorphic cohomology groups.

Replacing p_{X*} with the pushforward with compact supports functor $p_{X!}$ we define the **hypercohomology with compact support**

$$H_c^i(X; \mathcal{F}^\bullet) = H^i(p_{X!}I^\bullet(\mathcal{F})).$$

How does this definition compare to our earlier one of the Čech hypercohomology of a complex of sheaves?

Theorem 3.6.2 (Leray acyclic covering theorem, see e.g. Kashiwara and Schapira [97, Prop. 2.8.5]). *Let $\mathcal{F}^\bullet$ be a complex of sheaves on a topological space X. Suppose $\mathcal{U}$ is an open cover of X for which*

$$H^i(U_1 \cap \ldots \cap U_n; \mathcal{F}^\bullet) = 0 \qquad \forall\, U_1, \ldots, U_n \in \mathcal{U}, n > 0 \text{ and } i > 0.$$

Then for each i there is a natural isomorphism

$$\check{H}^i(X; \mathcal{F}^\bullet) \cong H^i(X; \mathcal{F}^\bullet).$$

Combining this with Proposition 3.2.3 we can show that for the constant sheaf $\mathbb{F}_X$ on a triangulable space X we have isomorphisms

$$H^i(X; \mathbb{F}_X) \cong \check{H}^i(X; \mathbb{F}_X) \cong H^i_{\mathrm{sing}}(X) \cong H^i_{\mathrm{simp}}(X);$$

for 'good' spaces all our definitions of cohomology agree.

Remark 3.6.3. It is easy to check that $H^0(X, \mathcal{F}) = \mathcal{F}(X) = \check{H}^0(X, \mathcal{F})$.

3.7   Derived categories

The derived category of sheaves on a space X is the natural algebraic framework for sheaf cohomology. It will allow us to give a clean formulation of the topological invariance of intersection cohomology in Chapter 7 and provides the setting for the discussion of perverse sheaves in Chapter 8. We will assume (very) basic familiarity with the notions of category theory — Mac Lane [123] is a good reference.

   The idea of the derived category is to treat complexes of sheaves and their resolutions on an equal footing, i.e. to treat quasi-isomorphisms as isomorphisms.

Definition 3.7.1. The objects of the bounded derived category $D^b(X)$ of sheaves on a topological space X are bounded complexes of sheaves of vector spaces over $\mathbb{F}$. A map in the derived category from a complex $\mathcal{F}^\bullet$ to a complex $\mathcal{G}^\bullet$ is given by an equivalence class of diagrams

$$\mathcal{F}^\bullet \xleftarrow{\phi^\bullet} \mathcal{E}^\bullet \xrightarrow{\psi^\bullet} \mathcal{G}^\bullet$$

in which $\phi^\bullet$ and $\psi^\bullet$ are cochain maps and $\phi^\bullet$ is a quasi-isomorphism. Two such diagrams $\mathcal{F}^\bullet \leftarrow \mathcal{E}_i^\bullet \to \mathcal{G}^\bullet$ for $i = 1, 2$ are equivalent if there is a commutative diagram of cochain maps

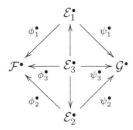

in which $\phi_3^\bullet$ is also a quasi-isomorphism. We write $\mathrm{Hom}_{D^b(X)}(\mathcal{F}^\bullet, \mathcal{G}^\bullet)$ for the set of maps from $\mathcal{F}^\bullet$ to $\mathcal{G}^\bullet$. It is easily seen to be a vector space over $\mathbb{F}$.

   One should think of this definition as analogous to the formation of a field of fractions of an integral domain by formally inverting elements and imposing an equivalence relation. In the case of the derived category we are formally inverting quasi-isomorphisms.

Exercise 3.7.2. Show that if $\phi^\bullet : \mathcal{F}^\bullet \to \mathcal{G}^\bullet$ is a quasi-isomorphism of bounded complexes of sheaves then it has an inverse in the bounded derived category $D^b(X)$.

Remark 3.7.3. There is nothing 'sheaf-y' about this construction; we could define the bounded derived category of any Abelian category, e.g. modules over a ring, in an analogous way.

The bounded derived category of sheaves on a topological space X satisfies the first three conditions to be an Abelian category (see page 34) but not the last two; there is no way to construct a kernel, image or cokernel of a map. In particular we cannot make sense of the concept of a short exact sequence in $D^b(X)$. There is however a substitute for this concept called a *distinguished triangle* which we now explain.

Definition 3.7.4. The **shifted complex** $\mathcal{F}^\bullet[1]$ of a complex $\mathcal{F}^\bullet$ of sheaves on X is defined by $(\mathcal{F}[1])^i = \mathcal{F}^{i-1}$ with the differential $-d$.

Suppose $\phi^\bullet : \mathcal{F}^\bullet \to \mathcal{G}^\bullet$ is a cochain map of complexes of sheaves on X. The **mapping cone** $\mathrm{Cone}^\bullet(\phi)$ of $\phi^\bullet$ is the complex with $\mathrm{Cone}^i(\phi) = \mathcal{F}^{i+1} \oplus \mathcal{G}^i$ and differential

$$\cdots \longrightarrow \mathcal{F}^{i+1} \oplus \mathcal{G}^i \xrightarrow{\begin{pmatrix} -d & 0 \\ \phi^{i+1} & d \end{pmatrix}} \mathcal{F}^{i+2} \oplus \mathcal{G}^{i+1} \longrightarrow \cdots$$

(This is a complex:

$$\begin{pmatrix} -d & 0 \\ \phi^{i+1} & d \end{pmatrix} \begin{pmatrix} -d & 0 \\ \phi^i & d \end{pmatrix} = \begin{pmatrix} d^2 & 0 \\ d \circ \phi^i - \phi^{i+1} \circ d & d^2 \end{pmatrix} = \begin{pmatrix} 0 & 0 \\ 0 & 0 \end{pmatrix}$$

because $\phi^\bullet$ is a cochain map, i.e. commutes with the differentials.)

Definition 3.7.5. A **standard triangle** is a sequence

$$\mathcal{F}^\bullet \xrightarrow{\phi^\bullet} \mathcal{G}^\bullet \xrightarrow{\begin{pmatrix} 0 \\ 1 \end{pmatrix}} \mathrm{Cone}^\bullet(\phi) \xrightarrow{(1\,0)} \mathcal{F}^\bullet[1]$$

of bounded complexes and cochain maps. Note that this can be extended to the left and to the right. A **distinguished triangle** is a sequence of bounded complexes and maps in the derived category

$$\mathcal{E}^\bullet \xrightarrow{\phi^\bullet} \mathcal{F}^\bullet \xrightarrow{\psi^\bullet} \mathcal{G}^\bullet \xrightarrow{\chi^\bullet} \mathcal{E}^\bullet[1]$$

which is isomorphic to a standard triangle in the sense that there is a commutative diagram

$$
\begin{array}{ccccccc}
\mathcal{A}^\bullet & \longrightarrow & \mathcal{B}^\bullet & \longrightarrow & \mathcal{C}^\bullet & \longrightarrow & \mathcal{A}^\bullet[1] \\
\downarrow & & \downarrow & & \downarrow & & \downarrow \\
\mathcal{E}^\bullet & \longrightarrow & \mathcal{F}^\bullet & \longrightarrow & \mathcal{G}^\bullet & \longrightarrow & \mathcal{E}^\bullet[1]
\end{array}
$$

in the derived category whose vertical maps are isomorphisms and whose top row is a standard triangle.

Example 3.7.6. We identify a sheaf on X with the complex of sheaves with that sheaf in degree zero and 0 in all other degrees. With this identification, sheaf maps correspond to cochain maps of sheaves in an obvious way.

If $0 \longrightarrow \mathcal{E} \overset{\phi}{\longrightarrow} \mathcal{F} \overset{\psi}{\longrightarrow} \mathcal{G} \longrightarrow 0$ is a short exact sequence of sheaves then $\mathrm{Cone}^\bullet(\phi)$ is the complex

$$0 \longrightarrow \mathcal{E} \overset{\phi}{\longrightarrow} \mathcal{F} \longrightarrow 0$$

where $\mathcal{F}$ is in degree zero. The cochain map

$$
\begin{array}{ccccccc}
0 & \longrightarrow & \mathcal{E} & \overset{\phi}{\longrightarrow} & \mathcal{F} & \longrightarrow & 0 \\
& & \downarrow & & \downarrow{\scriptstyle \psi} & & \downarrow \\
0 & \longrightarrow & 0 & \longrightarrow & \mathcal{G} & \longrightarrow & 0
\end{array}
\tag{3.6}
$$

from $\mathrm{Cone}^\bullet(\phi)$ to the complex with $\mathcal{G}$ in degree zero (and 0 in all other degrees) is a quasi-isomorphism and so, we have a distinguished triangle

$$\mathcal{E} \overset{\phi}{\longrightarrow} \mathcal{F} \overset{\psi}{\longrightarrow} \mathcal{G} \overset{\chi}{\longrightarrow} \mathcal{E}[1]$$

where χ is the (vertical) composite

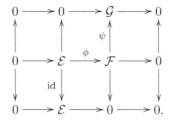

Thus each short exact sequence of sheaves determines a distinguished triangle.

Remark 3.7.7. The derived category together with the collection of distinguished triangles forms a **triangulated category** — see Gelfand and Manin [63, Ch. 5] for more detail.

The following proposition clarifies the sense in which distinguished triangles are a substitute for short exact sequences.

Proposition 3.7.8 (Gelfand and Manin [63, Ch. 5, §1.6]). *If* $\mathcal{E}^\bullet \overset{\phi^\bullet}{\longrightarrow} \mathcal{F}^\bullet \overset{\psi^\bullet}{\longrightarrow} \mathcal{G}^\bullet \overset{\chi^\bullet}{\longrightarrow} \mathcal{E}^\bullet[1]$ *is a distinguished triangle in* $D^b(X)$ *then there is an induced exact sequence of cohomology sheaves*

$$\cdots \to \mathcal{H}^i(\mathcal{E}) \to \mathcal{H}^i(\mathcal{F}) \to \mathcal{H}^i(\mathcal{G}) \to \mathcal{H}^{i+1}(\mathcal{E}) \to \cdots$$

In particular a cochain map $\phi^\bullet$ *is a quasi-isomorphism if and only if*

$$\mathcal{H}^i(\mathrm{Cone}^\bullet(\phi)) = 0 \qquad \forall i$$

or equivalently if and only if $\mathrm{Cone}^\bullet(\phi)$ *is an exact complex.*

3.8   Right derived functors

Suppose $F : \mathrm{Sh}(X) \to \mathrm{Sh}(Y)$ is a left exact functor. We would like to induce a functor $D^b(X) \to D^b(Y)$, but the simplistic definition 'apply F term-wise to complexes and cochain maps' doesn't make sense: for a general map

$$\mathcal{F}^\bullet \xleftarrow{\phi^\bullet} \mathcal{E}^\bullet \xrightarrow{\psi^\bullet} \mathcal{G}^\bullet$$

in $D^b(X)$ the diagram

$$F\mathcal{F}^\bullet \xleftarrow{F\phi^\bullet} F\mathcal{E}^\bullet \xrightarrow{F\psi^\bullet} F\mathcal{G}^\bullet$$

need not represent a map in $D^b(Y)$ because there is no reason for $F\phi^\bullet$ to be a quasi-isomorphism. However, if $\phi^\bullet : \mathcal{I}^\bullet \to \mathcal{J}^\bullet$ is a quasi-isomorphism between complexes of *injective* sheaves then $\mathrm{Cone}^i(\phi) = \mathcal{I}^{i+1} \oplus \mathcal{J}^i$ is an injective sheaf. An easy corollary of Theorem 3.5.3 shows that F takes a bounded below exact sequence of injective sheaves to an exact sequence. Hence we have

$$\phi^\bullet \text{ a quasi-isomorphism} \quad \Rightarrow \quad \mathrm{Cone}^\bullet(\phi) \text{ exact}$$
$$\Rightarrow \quad F(\mathrm{Cone}^\bullet(\phi)) = \mathrm{Cone}^\bullet(F(\phi)) \text{ exact}$$
$$\Rightarrow \quad F(\phi^\bullet) \text{ a quasi-isomorphism.}$$

Recall that we fixed a functorial choice of injective resolution $\mathcal{E}^\bullet \mapsto I^\bullet(\mathcal{E})$ for each complex of sheaves on X. Another way of saying this is that we have chosen a functor

$$
\begin{aligned}
I : D^b(X) &\longrightarrow D^b(X) \\
\mathcal{E}^\bullet &\longmapsto I^\bullet(\mathcal{E})
\end{aligned}
$$

taking each complex to an isomorphic (in the derived category) complex of injective sheaves.

Definition 3.8.1. The **right derived functor**

$$RF : D^b(X) \to D^b(Y)$$

of a left exact functor $F : \mathrm{Sh}(X) \to \mathrm{Sh}(Y)$ is the composition $F \circ I$, i.e. $RF(\mathcal{E}^\bullet) = F(I^\bullet(\mathcal{E}))$. This is well defined on maps in the derived category because whenever $\phi^\bullet$ is a quasi-isomorphism then, by the above, so is $F(I^\bullet(\phi))$.

Example 3.8.2. The functor assigning the ith hypercohomology group to a complex $\mathcal{E}^\bullet$ can be described as the composite

$$D^b(X) \xrightarrow{Rp_{X*}} D^b(\mathrm{pt}) \xrightarrow{H^i} \mathbb{F}\text{-VS}$$

of the right derived functor of the pushforward to a point and the ith cohomology of a complex of vector spaces over $\mathbb{F}$. In particular, note that quasi-isomorphic complexes have isomorphic hypercohomology groups.

After this short introduction it should be apparent that the derived category provides a natural framework for discussing sheaf cohomology. In fact the relationship is even closer than we have revealed. With a little more homological algebra (see Gelfand and Manin [63, Ch. 4, §3 and Thm. 5.3]) one can show that there is a natural isomorphism

$$H^i(X; \mathcal{E}^\bullet) \cong \mathrm{Hom}_{D^b(X)}(\mathbb{F}_X, \mathcal{E}^\bullet[i])$$

between the ith hypercohomology group of a complex $\mathcal{E}^\bullet$ and maps in the derived category from the constant sheaf $\mathbb{F}_X$ to the ith shift $\mathcal{E}^\bullet[i]$ of the complex — maps in the derived category *are* cohomology classes (in a suitably generalised sense). This is a very pleasing description but we should also take it as a warning that all the subtleties of cohomology classes are inherited by maps in the derived category. In particular, and in contrast to maps of sheaves, maps in $D^b(X)$ can be 'non-local'. By this we mean that there can be non-zero maps in $D^b(X)$ whose restriction to all of the sets U_i in some open cover of X are zero as maps in $D^b(U_i)$.

3.9   Further reading

There are many treatments of the theory of sheaves and their cohomology. Godement [65] and Serre [160] are classic accounts. Hartshorne [79] and Iversen [84] are good references, the first from the viewpoint of algebraic geometry and the second from that of topology. Kashiwara–Schapira [97] is a comprehensive account of sheaves on manifolds, and contains the definitive treatment of the micro-local geometry of sheaves — a subject which we have neglected. Bott and Tu [26] contains a very accessible account of Čech cohomology explaining how it arises from the Mayer–Vietoris principle.

The two main aspects of derived functor cohomology, homological algebra and category theory, arose together: Cartan and Eilenberg [41] is a classic by the pioneers and Hilton and Stammbach [81] is a good standard reference. For a more modern treatment the reader should look at Weibel [177] or Gelfand and Manin's two books [64] and [63]. These include a wealth of applications in many areas of algebra, geometry and topology.

Chapter 4

The definition of intersection homology

In this chapter we (finally) come to the definition of intersection homology.

4.1 Stratified spaces and pseudomanifolds

Intersection homology can be defined for a wide class of singular spaces called topological pseudomanifolds, which we now define.

If L is a compact Hausdorff topological space then the **open cone** $C(L)$ on L is the result of identifying the subset $L \times \{0\}$ of $L \times [0, 1)$ to a single point (called the vertex of the cone).

Definition 4.1.1. We define a topologically stratified space inductively on dimension. A 0-dimensional stratified space X is a countable set with the discrete topology. For $m > 0$ an m-dimensional **topologically stratified space** is a para-compact Hausdorff topological space X equipped with a filtration

$$X = X_m \supseteq X_{m-1} \supseteq \cdots \supseteq X_1 \supseteq X_0$$

of X by closed subsets X_j such that if $x \in X_j - X_{j-1}$ there exists a neighbourhood N_x of x in X, a compact $(m-j-1)$-dimensional topologically stratified space L with filtration

$$L = L_{m-j-1} \supseteq \cdots \supseteq L_1 \supseteq L_0$$

and a homeomorphism

$$\phi \colon N_x \to \mathbb{R}^j \times C(L)$$

where $C(L)$ is the open cone on L, such that ϕ takes $N_x \cap X_{j+i+1}$ homeomorphically onto

$$\mathbb{R}^j \times C(L_i) \subseteq \mathbb{R}^j \times C(L)$$

for $m - j - 1 \geq i \geq 0$, and ϕ takes $N_x \cap X_j$ homeomorphically onto

$$\mathbb{R}^j \times \{\text{vertex of } C(L)\}.$$

In particular this guarantees that the subset $X_j - X_{j-1}$ is a topological manifold of dimension j. The **strata** of X are the connected components of these manifolds. Up to homeomorphism, the space L only depends on the stratum in which the point x lies. It is referred to as the **link** of the stratum.

Definition 4.1.2. A **topological pseudomanifold** of dimension m is a para-compact Hausdorff topological space X which possesses a topological stratification such that

$$X_{m-1} = X_{m-2}$$

and $X - X_{m-1}$ is dense in X. (Note that the stratification is not part of the data; topological pseudomanifolds are *stratifiable* not *stratified*.)

Examples 4.1.3.   1. Any manifold X is a topological pseudomanifold.

2. The open cone on a manifold of dimension ≥ 1 is always a pseudomanifold (but, for example, the open cone on three points is *not* a pseudomanifold).

3. Two spheres joined at a point and a pinched torus are pseudomanifolds

but a torus with a spanning disc across the central hole

is not.

4. Later in this chapter we will see that complex quasi-projective varieties are pseudomanifolds.

We say an m-dimensional pseudomanifold is **irreducible** if $X_m - X_{m-2}$ is connected, in which case $H_m(X; \mathbb{Z})$ is either $\mathbb{Z}$ or 0. If it is $\mathbb{Z}$ then we say X is **orientable** and a choice of generator for $H_m(X; \mathbb{Z})$ is an **orientation**.

4.2 Simplicial intersection homology

For the first definition of intersection homology which we shall give we actually need X to be more than a topological pseudomanifold with filtration X_j as above. We also require X to have a triangulation which is compatible with the filtration (i.e. each X_j is a union of simplices). X is then a **piecewise-linear pseudomanifold**.

Remark 4.2.1. We can characterise piecewise-linear pseudomanifolds more simply as follows. The geometric realisation of a simplicial complex K is an m-dimensional piecewise-linear pseudomanifold if and only if

1. every simplex is a face of an m-simplex;

2. every $(m-1)$-simplex is a face of precisely two m-simplices.

Note that not all topological pseudomanifolds are piecewise-linear. Indeed, there are even examples of topological manifolds which cannot be triangulated, see Ranicki *et al.* [145].

As in Chapter 2 we will work with coefficients in a field $\mathbb{F}$.

Let $T\colon |N| \to X$ be a triangulation of X compatible with the stratification. Recall that $C_i^T(X)$ is the space of all (finite) simplicial i-chains of X with respect to T.

Definition 4.2.2. The **support** $|\xi|$ of a simplicial i-chain $\xi = \sum_{\sigma \in N^{(i)}} \xi_\sigma \sigma$ is given by

$$|\xi| = \bigcup_{\xi_\sigma \neq 0} T(\sigma).$$

We are going to define a subspace $IC_i^T(X)$ of $C_i^T(X)$ whose elements will be those i-chains ξ such that the intersection of $|\xi|$ and X_j is 'not too big' for each j. To make 'not too big' precise we need the concept of a perversity.

Definition 4.2.3. A **perversity** is a function $p\colon \{2,\ldots,m\} \to \mathbb{N}$ such that $p(2) = 0$ and $p(i+1) = p(i)$ or $p(i)+1$.

Examples 4.2.4.   1. The zero perversity is $i \mapsto 0$ for $2 \leq i \leq m$.

2. The top perversity is $t(i) = i - 2$.

3. If p is a perversity, the **complementary perversity** is

$$i \mapsto t(i) - p(i) = i - p(i) - 2.$$

4. The lower middle perversity is $m(i) = \lfloor i/2 \rfloor - 1$ and the upper middle perversity is its complement $n(i) = \lceil i/2 \rceil - 1$.

We fix a perversity p.

Definition 4.2.5. We say an i-chain $\xi \in C_i^T(X)$ is p**-allowable** if

$$\dim_{\mathbb{R}} |\xi| \cap X_{m-k} \le i - k + p(k).$$

Note that by convention the empty set has dimension $-\infty$. Let $I^pC_i^T(X)$ be the subspace of $C_i^T(X)$ consisting of all those i-chains $\xi \in C_i^T(X)$ such that ξ is a p-allowable i-chain and $\partial\xi$ a p-allowable $(i-1)$-chain.

Remark 4.2.6. Since the triangulation T is compatible with the stratification, the intersection $|\xi| \cap X_{m-k}$ is a union of simplices and hence has a well-defined dimension. We can make an analogous definition for any version of homology with chains for which this holds, for example for semi-analytic chains (see e.g. Hardt [78]).

Since X_{m-k} has codimension k an i-chain is dimensionally transverse to it if $\dim_{\mathbb{R}} |\xi| \cap X_{m-k} \le i - k$. Thus the value $p(k)$ of the perversity tells us how far from dimensional tranversality to codimension k strata an i-chain is allowed to be.

It is easy to check that if T' is a refinement of the triangulation T then the induced map

$$C_i^T(X) \to C_i^{T'}(X)$$

sends a chain $\xi \in C_i^T(X)$ to a chain with the same support as ξ. Hence it restricts to maps

$$I^pC_i^T(X) \to I^pC_i^{T'}(X).$$

Definition 4.2.7. The space $I^pC_i(X)$ of **piecewise-linear intersection i-chains** is the limit of the $I^pC_i^T(X)$ over all triangulations T of X compatible with the stratification.

Thus a piecewise-linear intersection i-chain is represented by an element of $I^pC_i^T(X)$ for some T, and

$$\eta \in I^pC_i^T(X) \quad \text{and} \quad \zeta \in I^pC_i^{T'}(X)$$

represent the same element of $I^pC_i(X)$ if and only if there is a common refinement T'' of T and T', compatible with the stratification, such that η and ζ induce the same element of

$$I^pC_i^{T''}(X).$$

It is easy to check from the definition, and the fact that $\partial^2 = 0$, that the boundary maps $\partial\colon C_i(X) \to C_{i-1}(X)$ induce boundary maps from $I^pC_i(X)$ to $I^pC_{i-1}(X)$.

Definition 4.2.8. The ith **intersection homology group** of X with perversity p is

$$I^pH_i(X) = \frac{\ker \partial\colon I^pC_i(X) \to I^pC_{i-1}(X)}{\operatorname{im} \partial\colon I^pC_{i+1}(X) \to I^pC_i(X)}.$$

The triangulation dependent group $I^pH_i^T(X)$ and the intersection cohomology groups $I^pH^i(X)$ and $I^pH_T^i(X)$ are defined similarly.

Remark 4.2.9. If T is any triangulation of a space X then $H_*(X) \cong H_*^T(X)$ (see e.g. Hatcher [80, §2.1]). This is not true for intersection homology, even for triangulations compatible with a given stratification. However, if the triangulation is **flag-like**, meaning that for each i the intersection of any simplex σ with the closure $\overline{X_i}$ is a single face of σ, then $I^p H_*(X) \cong I^p H_*^T(X)$ for any perversity p (Goresky and MacPherson [73]). Thus intersection homology is computable from a flag-like triangulation.

Of course a priori $I^p H_i(X)$ depends on the choice of stratification of X. We shall see later that in fact it is independent of this choice (Goresky and MacPherson [70, §4]).

Remark 4.2.10. The attentive reader will have noticed that the definition of intersection homology depends only on the filtration by closed subsets and that we have not used the locally cone-like structure of a topological pseudomanifold. Thus we can extend the above definitions and assign intersection homology groups to any filtered space (even if the 'strata' are not manifolds). However, these groups are not in general invariants of the underlying space but only of the filtration. The extra structure possessed by pseudomanifolds ensures that the intersection homology groups are topological invariants — see Chapter 7.

The lower and upper middle perversities will be the most important for us and we now make an arbitrary choice to favour the lower middle perversity m and, for simplicity of notation, put

$$IC_i(X) = I^m C_i(X), \quad IH_i(X) = I^m H_i(X) \tag{4.1}$$

etc.

Remark 4.2.11. We can also define intersection homology groups with **closed support** by using locally finite intersection chains. The definitions are entirely analogous but with $C_i^T(X)$ replaced by $C_i^T((X))$ throughout. We denote the group of locally finite intersection chains by $IC_i((X))$ and the intersection homology groups with closed support by $IH_i^{cl}(X)$.

The definitions of intersection homology given in the literature are inconsistent, and often the groups $IH_i^{cl}(X)$ are called the intersection homology groups of X instead of the groups $IH_i(X)$ defined at 4.2.8. This is because the groups $IH_i^{cl}(X)$ fit better with the sheaf-theoretic approach to intersection homology (see Chapter 7) although the groups $IH_i(X)$ fit better with the classical homology theory. Of course when X is compact there is a natural identification

$$IH_i(X) \cong IH_i^{cl}(X)$$

so it does not matter which definition is used.

4.3   Singular intersection homology

Intersection homology can also be defined from the point of view of singular homology theory. This approach is due to King [110] and has the advantage that we do not require a triangulation and so it works for any topological pseudomanifold.

Suppose X is a topological m-pseudomanifold and that we have chosen a stratification

$$X = X_m \supseteq X_{m-1} = X_{m-2} \supseteq \cdots \supseteq X_1 \supseteq X_0.$$

As before we fix a perversity $p : \{2, \ldots, m\} \to \mathbb{N}$.

Definition 4.3.1. Let Δ_i be the standard i-simplex in $\mathbb{R}^{i+1}$. The j-**skeleton** of Δ_i is the set of j-subsimplices. We say a singular i-simplex in X, i.e. a continuous map $\sigma : \Delta_i \to X$, is p-**allowable** if

$$\sigma^{-1}(X_{m-k} - X_{m-k-1}) \subset (i - k + p(k))\text{-skeleton of } \Delta_i$$

for $k \geq 2$. A singular i-chain is p-allowable if it is a formal linear combination of p-allowable singular simplices. Note that this definition of allowability is consistent with our earlier one in the simplicial theory.

The subspace $I^p S_i(X) \subset S_i(X)$ of the singular i-chains consists of those p-allowable chains with p-allowable boundary. This clearly forms a subcomplex and the **perversity** p **singular intersection homology groups** are the homology groups of this complex. When X is a piecewise-linear pseudomanifold these are canonically isomorphic to the previously defined simplicial intersection homology groups, see King [110], and so we denote them in the same way by $I^p H_i(X)$ etc. As before we privilege the lower middle perversity and simply write $IH_i(X)$ for $I^m H_i(X)$.

Remark 4.3.2. We can also define singular intersection homology with closed supports by considering p-allowable locally finite singular chains cf. §2.3.

4.4   Simple examples of intersection homology

The simplest case is when X is a topological manifold. Then $I^p H_*(X) = H_*(X)$ for any perversity p. The next simplest case is when X is an irreducible topological pseudomanifold with one isolated singularity x, so that $X - \{x\}$ is a manifold. For simplicity we assume that the dimension of X is even, say $2k$ (the odd dimensional case is similar). The filtration

$$X = X_{2k} \supseteq X_{2k-1} \supseteq \cdots \supseteq X_0$$

with $X_j = \{x\}$ if $0 \leq j < 2k$ defines a stratification of X.

Proposition 4.4.1. *The lower middle perversity intersection homology is given by*

$$IH_i(X) = \begin{cases} H_i(X) & i > k \\ \operatorname{im}(H_i(X - \{x\}) \to H_i(X)) & i = k \\ H_i(X - \{x\}) & i < k. \end{cases}$$

Proof. For simplicity we assume X is compatibly triangulated. The singular computation is similar. The group $IC_i(X)$ of simplicial intersection i-chains is given by

$$\{\xi \in C_i(X) | \dim |\xi| \cap \{x\} \le i - m - 1, \dim |\partial \xi| \cap \{x\} \le i - k - 2\}. \quad (4.2)$$

Hence if $i \le k$ then $IC_i(X) = IC_i(X - \{x\}) = C_i(X - \{x\})$, whereas if $i \ge k + 2$ then $IC_i(X) = C_i(X)$. Hence $IH_i(X) = H_i(X - \{x\})$ if $i \le k - 1$ and $IH_i(X) = H_i(X)$ if $i \ge k + 2$. Moreover

$$\ker(\partial: IC_{k+1}(X) \to IC_k(X)) = \ker(\partial: C_{k+1}(X) \to C_k(X))$$

so $IH_{k+1}(X) \cong H_{k+1}(X)$. Finally

$$\partial(IC_{k+1}(X)) = (\partial C_{k+1}(X)) \cap IC_k(X)$$

and $IC_k(X) = C_k(X - \{x\})$ so $IH_k(X) \cong \operatorname{im}(H_k(X - \{x\}) \to H_k(X))$. $\square$

As specific examples consider X_1, the union of two spheres at a point:

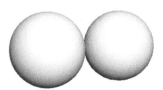

and X_2, the result of collapsing one of the standard generators of the first homology of the torus:

We have

$$IH_i(X_1) = \begin{cases} \mathbb{F} \oplus \mathbb{F} & i = 0 \\ 0 & i = 1 \\ \mathbb{F} \oplus \mathbb{F} & i = 2 \end{cases} \quad \text{and} \quad IH_i(X_2) = \begin{cases} \mathbb{F} & i = 0 \\ 0 & i = 1 \\ \mathbb{F} & i = 2. \end{cases} \quad (4.3)$$

Remark 4.4.2. Another way of viewing this result is that a pseudomanifold with an isolated singularity can be formed by coning off a manifold with boundary (the boundary is the link of the singular point). Suppose $(M, \partial M)$ is a $2k$-manifold with boundary and let

$$\widehat{M} = M \cup_{\partial M} C(\partial M)$$

be the space formed by glueing the cone $C(\partial M)$ onto the boundary. It follows from the long exact sequence

$$\cdots \to H_i(C(\partial M)) \to H_i(\widehat{M}) \to H_i(\widehat{M}, C(\partial M)) \to \cdots$$

and excision that $H_i(\widehat{M}) \cong H_i(M, \partial M)$ for $i > 0$. By homotopy invariance

$$H_i(\widehat{M} - \text{vertex}) \cong H_i(M).$$

Thus we can rewrite Proposition 4.4.1 as

$$IH_i(\widehat{M}) = \begin{cases} H_i(M, \partial M) & i > k \\ \text{im}\,(H_i(M) \to H_i(M, \partial M)) & i = k \\ H_i(M) & i < k. \end{cases}$$

4.5   Normalisations

Let X be a topological pseudomanifold with filtration

$$X = X_m \supseteq X_{m-1} = X_{m-2} \supseteq \cdots \supseteq X_0.$$

Then X is called (topologically) **normal** if every $x \in X$ has an open neighbourhood U in X such that $U - X_{m-2}$ is connected. Note that X is normal if and only if the link of any stratum is connected. Any manifold is normal. Furthermore any topological pseudomanifold X has a **normalisation** $\pi : \tilde{X} \to X$. Here π is a continuous surjection from a normal topological pseudomanifold $\tilde{X}$ onto X which is a homeomorphism onto X_{m-2}. This is constructed as follows. Recall that each point $x \in X$ has a neighbourhood N_x which is homeomorphic to $\mathbb{R}^j \times C(L)$ where L is the link of the stratum in which x lies. Suppose L has connected components $L_0, \ldots, L_n$. Define $\tilde{N}_x = \bigcup_i \mathbb{R}^j \times C(L_i)$ and let π be the obvious surjection onto N_x. Then put

$$\tilde{X} = \left(\bigcup_{x \in X} \tilde{N}_x \right) / R$$

where the equivalence relation R is given by pRq if $p, q \in \pi^{-1}(X - X_{m-2})$ and $\pi(p) = \pi(q)$.

Proposition 4.5.1 (Goresky and MacPherson [68, §4.3]). *Let X be a topological pseudomanifold of dimension m. If X is topologically normal then there are canonical isomorphisms*

$$I^t H_i(X) \cong H_i(X) \quad and \quad I^0 H_i(X) \cong H^{m-i}(X)$$

where $t : j \mapsto j - 2$ is the top perversity and $0 : j \mapsto 0$ is the zero perversity.

Proposition 4.5.2 (Goresky and MacPherson [68, §4.2]). *If $\pi : \tilde{X} \to X$ is a normalisation of X then there is a natural isomorphism*

$$I^p H_i(\tilde{X}) \cong I^p H_i(X)$$

for any perversity p.

Example 4.5.3. Consider the pseudomanifolds X_1 and X_2 defined in the previous section. The normalisation of X_1 is the disjoint union of two copies of S^2 and the normalisation of X_2 is S^2. This fits with (4.3).

4.6 Relative groups and the Mayer–Vietoris sequence

Suppose that $U \subseteq X$ is an open subset of a topological pseudomanifold X. Composition with the inclusion $U \hookrightarrow X$ yields natural inclusions

$$I^p S_i(U) \to I^p S_i(X)$$

which commute with the boundary maps. Thus there is an induced complex

$$I^p S_*(X, U) = \frac{I^p S_*(X)}{I^p S_*(U)}$$

whose homology groups are the **relative intersection homology groups** of the pair (X, U). Just as for ordinary homology there is a long exact sequence

$$\cdots \to I^p H_i(U) \to I^p H_i(X) \to I^p H_i(X, U) \to I^p H_{i-1}(U) \to \cdots \quad (4.4)$$

(Goresky and MacPherson [71, §1.3]).

Warning 2. The ordinary relative homology groups $H_i(X, A)$ are defined for any subset A of X, but this fails for intersection homology.

The relative groups also have the following excisive property

Proposition 4.6.1 (Goresky and MacPherson [71, §1.5]). *Suppose U is an open subset of a topological pseudomanifold X and $A \subset U$ a closed subset of U such that $X - A$ (and hence $U - A$) is still a topological pseudomanifold. Then there are natural isomorphisms*

$$I^p H_i(X, U) \cong I^p H_i(X - A, U - A).$$

The easiest proof of this will follow from the sheaf-theoretic treatment of intersection homology we give in Chapter 7. Alternatively we can adapt the proof of the analogous result in simplicial or singular homology.

We also have Mayer–Vietoris sequences for intersection homology: whenever we express X as a union $U \cup V$ of two open sets then the usual proof for ordinary homology can be adapted to show that there is a long exact sequence

$$\cdots \to IH_i(U \cap V) \to IH_i(U) \oplus IH_i(V) \to IH_i(X) \to IH_{i-1}(U \cap V) \to \cdots .$$
(4.5)

4.7   The intersection homology of a cone

An m-dimensional manifold is a topological space which is locally modelled on $\mathbb{R}^m$. The key homological calculations for manifold theory are

$$H_i(\mathbb{R}^m) = \begin{cases} \mathbb{F} & i = 0 \\ 0 & i \neq 0 \end{cases} \quad \text{and} \quad H_i(\mathbb{R}^m, \mathbb{R}^m - \{0\}) = \begin{cases} \mathbb{F} & i = m \\ 0 & i \neq m. \end{cases} \quad (4.6)$$

We are interested in m-dimensional topological pseudomanifolds, spaces which are locally modelled on the open cone on a compact topological pseudomanifold of dimension $m - 1$. Thus the key calculations for us will be the intersection homology of an open cone and the intersection homology of the cone relative to the cone less the vertex.

We will use the following simple version of the Künneth theorem. The proof is an easy adaptation of the proof for singular homology (see Goresky and MacPherson [71, §1.6]).

Proposition 4.7.1 (cf. King [110, Lem. 3]). *If X is a topological pseudomanifold then so is $X \times (0,1)$ and, for any perversity p, we have*

$$I^p H_i(X \times (0,1)) \cong I^p H_i(X).$$

Suppose X is a topological pseudomanifold of dimension $m \geq 1$ with a given stratification

$$X = X_m \supseteq X_{m-2} \supseteq \cdots \supseteq X_0.$$

Then the open cone $C(X)$ is also a topological pseudomanifold as it is naturally stratified by

$$C(X) = C(X_m) \supseteq C(X_{m-2}) \supseteq \cdots \supseteq C(X_0) \supseteq \{v\}.$$

where v is the vertex of the cone.

Proposition 4.7.2. *Suppose X is a compact topological pseudomanifold of dimension $m \geq 1$. Then, for a perversity p,*

$$I^p H_i(C(X)) \cong \begin{cases} I^p H_i(X) & i < m - p(m+1) \\ 0 & otherwise \end{cases}$$

and

$$I^p H_i\left(C(X), C(X) - \{v\}\right) \cong \begin{cases} 0 & i \leq m - p(m+1) \\ I^p H_{i-1}(X) & otherwise. \end{cases}$$

Proof. The vertex of the cone has codimension $m+1$ and so for $i \leq m - p(m+1)$ an allowable singular i-simplex cannot meet the vertex because

$$i - (m+1) + p(m+1) < 0.$$

Hence $I^p S_i\left(C(X)\right) = I^p S_i\left(C(X) - \{v\}\right)$ for $i \leq m - p(m+1)$ and so, for $i < m - p(m+1)$,

$$\begin{aligned} I^p H_i\left(C(X)\right) &\cong I^p H_i\left(C(X) - \{v\}\right) \\ &\cong I^p H_i\left(X \times (0,1)\right) \\ &\cong I^p H_i(X) \end{aligned}$$

by the Künneth theorem.

Now suppose $\sigma : \Delta_i \to X$ is an allowable singular i-simplex for some $i \geq m - p(m+1)$. Then, representing a point in Δ_{i+1} by a pair $[s,t]$ with $s \in \Delta_i$ and $t \in [0,1]$, we can define a singular $(i+1)$-simplex

$$c\sigma : [s,t] \mapsto t\sigma(s) \in C(X)$$

(which should be thought of as the cone on σ). Suppose S_k is a stratum of $C(X)$ of codimension k, i.e. if $k < m+1$ then S_k is a connected component of $(X_k - X_{k-1}) \times (0,1)$ and if $k = m+1$ then S_k is the vertex. The simplex $c\sigma$ is allowable because

- if $\sigma^{-1}(S_k) \neq \emptyset$ then $\sigma^{-1}(S_k)$ is contained in the $i - k + p(k)$ skeleton of Δ_i and so $(c\sigma)^{-1}(S_k)$ is contained in the $(i+1) - k + p(k)$ skeleton of Δ_{i+1};

- and if $\sigma^{-1}(S_k) = \emptyset$ then $(c\sigma)^{-1}(S_k) = \emptyset$ *unless* S_k is the vertex in which case $(c\sigma)^{-1}(S_k)$ is contained in the 0-skeleton of Δ_{i+1}, which is allowed since $i + 1 > m - p(m+1)$.

Extending the definition of c linearly to chains, and taking careful account of signs, $\partial(c\sigma) + c(\partial\sigma) = \sigma$. Hence if $\xi \in I^p C_i\left(C(X)\right)$ has $\partial\xi = 0$ then $\xi = \partial(c\xi)$ is a boundary and so

$$I^p H_i\left(C(X)\right) = 0$$

for $i \geq m - p(m+1)$.

The computation of the relative group $I^p H_i\left(C(X), C(X) - \{v\}\right)$ now follows from the long exact sequence

$$\cdots \to I^p H_i\left(C(X) - \{v\}\right) \to I^p H_i\left(C(X)\right) \to I^p H_i\left(C(X), C(X) - \{v\}\right) \to \cdots$$

$\square$

Exercise 4.7.3. Let X be a $(2k-1)$-dimensionsal pseudomanifold. Use the cone calculation and Mayer–Vietoris to compute the intersection homology of the suspension $\mathrm{Susp}(X) = (X \times [-1,1]) / (X \times \{-1,1\})$. In particular show that if $\dim X = 2k$ then the lower middle perversity groups are

$$IH_i(\mathrm{Susp}(X)) = \begin{cases} IH_i(X) & i < k \\ 0 & i = k \\ IH_{i-1}(X) & i > k. \end{cases}$$

(Intuitively, in low dimensions allowable cycles must avoid the suspension points and in high dimensions the only allowable cycles which cannot be 'coned off' are the suspensions of allowable cycles in X.)

An analogous calculation for the simplicial intersection homology of a cone can be found in Goresky and MacPherson [70, §2.4]. It is worth dwelling for a moment on this calculation and its implications. Firstly note that $\mathbb{R}^m$ is homeomorphic to the open cone on S^{m-1} and so, in particular, we recover (4.6).

Since a topological pseudomanifold is locally modelled on a product of an open cone and $\mathbb{R}^n$ we can (with repeated applications of the Künneth theorem) compute the intersection homology $I^p H_i(N_x)$ of a neighbourhood N_x of a point x, and also the relative group $I^p H_i(N_x, N_x - \{x\})$. These calculations will be central to our sheaf-theoretic treatment of intersection homology in Chapter 7.

We can give an intuitive description of these calculations as follows. In low dimensions allowable chains cannot meet the vertex so the intersection homology is that of the cone less the vertex. In high dimensions we can cone off chains to the vertex and so the intersection homology vanishes. The relative calculation is more easily visualised if we reinterpret it in terms of closed support intersection homology (this is the approach taken in Borel *et al.* [24, Ch. I]). There are natural maps

$$I^p S_i^{cl}(C(X)) \to I^p S_i(C(X)) / I^p S_i(C(X) - \{v\})$$

for each i which induce isomorphisms

$$I^p H_i^{cl}(C(X)) \cong I^p H_i(C(X), C(X) - \{v\}). \tag{4.7}$$

The closed support intersection homology vanishes in low dimensions because chains which do not meet the vertex can be 'coned off to ∞'. In high dimensions closed support intersection homology classes arise as the open cone on classes in the intersection homology of the cone less the vertex, which is why we see a dimension shift.

Both the intersection homology of $C(X)$ and the closed support intersection homology (or equivalently the corresponding relative group) arise entirely from the intersection homology of the link X of the vertex (and the intersection homology of the link together with the choice of perversity determine both). Since the link is a compact topological pseudomanifold of one lower dimension than the cone one can often use induction on dimension to prove results about intersection homology. An important example will occur when we discuss generalised Poincaré duality in Chapter 5.

The perversity can be viewed as a choice of assignment, according to dimension, of the intersection homology of the link between the intersection homology and closed support intersection homology of the cone — classes in dimension $< m - p(m+1)$ are assigned to the former and those of dimension $\geq m - p(m + 1)$ to the latter. The extreme cases are the zero perversity, which assigns all but the top group $I^pH_m(X)$ to $I^pH_*(C(X))$, and the top perversity, which assigns all but the bottom group $I^pH_0(X)$ to $I^pH_*^{cl}(C(X))$. This should de-mystify the, at first rather arbitrary seeming, definition of perversity. The lower and upper middle perversities make as even a split as possible.

This calculation also has implications for the functoriality of intersection homology, a subject which deserves its own section.

4.8 Functoriality of intersection homology

If $f : X \to Y$ is a continuous map then composition with f induces maps $S_i(X) \to S_i(Y)$ for each i which commute with the boundary maps, and hence induces a linear map $H_i(X) \to H_i(Y)$ on singular homology. However composition with f need not take a p-allowable simplex in $S_i(X)$ to a p-allowable simplex in $S_i(Y)$, as we can readily verify by considering the continuous map given by the inclusion of the vertex into an open cone. Thus, in contrast to ordinary homology, an arbitrary continuous map $f: X \to Y$ does not in general induce a homomorphism $f_*: I^pH_*(X) \to I^pH_*(Y)$. Furthermore, the inclusion of the vertex into an open cone is a homotopy equivalence, and yet we can easily construct examples in which the intersection homology of the cone is different from that of a point — intersection homology is not a homotopy invariant. What then can we say?

Definition 4.8.1. A continuous map $f : X \to Y$ between topologically stratified spaces is **stratum-preserving** if, for each stratum T of Y, the inverse image $f^{-1}(T)$ is a union of strata of X. Equivalently, the image $f(S)$ of each stratum S of X is contained within a single stratum of Y. A stratum-preserving map $f : X \to Y$ is **placid** if for each stratum T of Y, we have

$$\operatorname{codim} f^{-1}(T) \geq \operatorname{codim} T. \tag{4.8}$$

Exercise 4.8.2. Show that a placid map $f : X \to Y$ induces linear maps $f_* : I^pH_i(X) \cong I^pH_i(Y)$ on singular intersection homology groups.

If we further strengthen the conditions on the map then we can obtain a stratified version of homotopy invariance for intersection homology.

Definition 4.8.3. A stratum-preserving map $f : X \to Y$ between topologically stratified spaces is **codimension-preserving** if, for each stratum T of Y we have

$$\operatorname{codim} f^{-1}(T) = \operatorname{codim} T. \tag{4.9}$$

Definition 4.8.4. A **stratum-preserving homotopy equivalence** between topological pseudomanifolds X and Y with chosen stratifications $\{X_i\}$ and $\{Y_j\}$ is a pair of codimension-preserving maps $f : X \to Y$ and $g : Y \to X$ such that $g \circ f$ and $f \circ g$ are homotopic to the identity via codimension-preserving homotopies

$$h : X \times [0,1] \to X \quad \text{and} \quad k : Y \times [0,1] \to Y$$

respectively. Here the stratification of $X \times [0,1]$ is given by $\{X_i \times [0,1]\}$, and that of $Y \times [0,1]$ by $\{Y_j \times [0,1]\}$.

Proposition 4.8.5 (Friedman [60, Prop. 2.1]). *If $f : X \to Y$ is a stratum-preserving homotopy equivalence then composition with f induces isomorphisms*

$$f_* : I^p H_i(X) \cong I^p H_i(Y).$$

Example 4.8.6. The obvious inclusion of a topological pseudomanifold X into $X \times (0,1)$ is a stratum-preserving homotopy equivalence and so the simple Künneth theorem 4.7.1 is a corollary.

Stratum-preserving maps are the natural maps between filtered spaces. Our spaces are not just filtered but topologically stratified and there is a more rigid concept of map which is useful in some contexts.

Definition 4.8.7. A **stratified map** $f : X \to Y$ between topologically stratified spaces is a stratum-preserving map such that, for each stratum T of Y, the restriction $f : f^{-1}(T) \to T$ is a locally trivial fibre bundle whose fibre is a topologically stratified space. In other words, for each $y \in T$ there exists a neighbourhood N_y of y in T, a topologically stratified space F_y and a stratum-preserving homeomorphism

$$\phi_y : N_y \times F_y \to f^{-1}(N_y)$$

where $N_y \times F_y$ has the product stratification. (In fact, up to homeomorphism, F_y depends only on the stratum T and not on y.)

The above definitions of stratum-preserving maps etc. rely on having fixed topological stratifications for X and Y. However, we are often interested in the topology of the underlying spaces and not in the properties of a particular stratification. We could rewrite the definitions conditionally; for example a map $f : X \to Y$ is *placid* if topological stratifications of X and Y exist for which it is stratum-preserving and satisfies the condition (4.8). However, it should be noted that the composition of two placid maps in this sense need not be placid; we cannot necessarily choose topological stratifications for which both maps are simultaneously placid. In general, the problem of defining a suitable category of topological spaces and maps for which intersection homology is functorial is a difficult one.

We end this discussion more positively with an important result — intersection homology is a *topological invariant*. In other words any homeomorphism $f \colon X \to Y$ between topological pseudomanifolds will induce an isomorphism

$$f_* \colon IH_*(X) \to IH_*(Y).$$

In particular, the intersection homology groups are independent of the stratification used to define them, and, in the simplicial case, independent of the chosen triangulation. A sketch proof will appear in Chapter 7.

4.9 Homology with local coefficients

We can also define intersection homology groups with coefficients in a local system. Let X be a stratified space.

Definition 4.9.1. A **local system** $\mathcal{L}$ of finite dimensional vector spaces over $\mathbb{F}$ on X is given by data consisting of a finite dimensional vector space $\mathcal{L}_x$ for each $x \in X$ and an isomorphism

$$\varphi^* \colon \mathcal{L}_{\varphi(0)} \to \mathcal{L}_{\varphi(1)}$$

for any continuous path $\varphi \colon [0,1] \to X$ satisfying

1. $\varphi^* = \psi^*$ when φ and ψ are homotopic relative to fixed end points, and

2. $(\varphi \cdot \psi)^* = \psi^* \circ \varphi^*$ if $\varphi(1) = \psi(0)$ and $\varphi \cdot \psi$ is the composite path from $\varphi(0)$ to $\psi(1)$.

Remark 4.9.2. Equivalently (if X is connected) $\mathcal{L}$ is given by a locally constant sheaf, that is a sheaf on X which is locally isomorphic to a constant sheaf defined by a finite-dimensional vector space, or by a representation of the fundamental group $\pi_1(X)$ on a finite dimensional vector space, or alternatively by a vector bundle on X equipped with a flat connection.

A global section of $\mathcal{L}$ is given by a map

$$g \colon X \to \coprod_{x \in X} \mathcal{L}_x$$

such that $g(x) \in \mathcal{L}_x$ for all $x \in X$, and if φ is a path from x to y then

$$\varphi^*(g(x)) = g(y).$$

(Thinking of $\mathcal{L}$ as a representation of $\pi_1(X)$ a section is a generator of a 1-dimensional subrepresentation or, thinking of $\mathcal{L}$ as a vector bundle with flat connection, a flat section of the bundle.)

If $\sigma\colon \Delta_i \to X$ is a singular i-simplex, then the pulled back sheaf $\sigma^*\mathcal{L}$ on Δ_i is also locally constant and, since Δ_i is simply connected, it is in fact constant. So, if L_σ is the space of sections of $\sigma^*\mathcal{L}$ over σ then the restriction maps

$$\rho_p^\sigma\colon L_\sigma \to (\sigma^*\mathcal{L})_p = \mathcal{L}_{\sigma(p)}$$

are isomorphisms for all $p \in \sigma$. Moreover if τ is a face of σ and $p \in \tau$ the composition

$$\rho_\tau^\sigma = (\rho_p^\tau)^{-1} \circ \rho_p^\sigma \colon L_\sigma \to L_\tau$$

is independent of p. Let $S_i(X;\mathcal{L})$ be the vector space consisting of all formal finite linear combinations

$$\xi = \sum_\sigma \ell_\sigma \sigma$$

of singular i-simplices with $\ell_\sigma \in L_\sigma$. Define $\partial\colon S_i(X;\mathcal{L}) \to S_{i-1}(X;\mathcal{L})$ by

$$\partial(\xi) = \sum_{\tau \text{ face of } \sigma} \pm \rho_\tau^\sigma(\ell_\sigma)\tau$$

where the sign $\pm$ is defined as before depending on a fixed choice of orientations. The ith **homology group of X with coefficients in** $\mathcal{L}$ is by definition the quotient

$$H_i(X;\mathcal{L}) = \frac{\ker \partial\colon S_i(X;\mathcal{L}) \to S_{i-1}(X;\mathcal{L})}{\operatorname{im} \partial\colon S_{i+1}(X;\mathcal{L}) \to S_i(X;\mathcal{L})}. \tag{4.10}$$

Now suppose that X is a topological pseudomanifold with a fixed topological stratification

$$X = X_m \supseteq X_{m-2} \supseteq \cdots \supseteq X_0.$$

To make this procedure work for intersection homology we only need the local coefficient system $\mathcal{L}$ to be defined on the open subset $X - X_{m-2}$ of X, not on the whole of X. This is because the allowability conditions on intersection i-chains ξ mean that if the coefficient of ξ indexed by σ is non-zero then

$$\sigma^{-1}(X - X_{m-2}) \neq \emptyset$$

and similarly $\tau^{-1}(X - X_{m-2}) \neq \emptyset$ for any face τ of σ. Thus we can use this procedure to define the intersection homology groups $IH_i(X;\mathcal{L})$ of X with coefficients in $\mathcal{L}$ for any local coefficient system $\mathcal{L}$ on $X - X_{m-2}$.

4.10 Quasi-projective complex varieties

We end this chapter with a brief discussion of the intersection homology of complex varieties. This has various good properties, collectively known as the Kähler package.

A **complex affine variety** is a subset X of some $\mathbb{C}^N$ defined by the simultaneous vanishing of polynomial equations. A **complex projective variety** X is a subset

$$X \subseteq \mathbb{CP}^N = \frac{\mathbb{C}^{N+1} - \{0\}}{\mathbb{C} - \{0\}}$$

of some complex projective space $\mathbb{CP}^N$ which is defined by the vanishing of homogenous polynomial equations. A **quasi-projective complex variety** X is a subset of $\mathbb{CP}^N$ of the form

$$X = Z - Y$$

where Z and Y are projective subvarieties of $\mathbb{CP}^N$. That is, there exists homogeneous polynomials $f_1, ..., f_r$ and $g_1, ...g_s$ in $N+1$ variables such that a point $(x_0 : ... : x_N) \in \mathbb{CP}^N$ belongs to X if and only if $f_j(x_0, ..., x_N) = 0$ for all j such that $1 \leq j \leq r$, and $g_j(x_0, ..., x_N) \neq 0$ for some j such that $1 \leq j \leq s$.

For example $\mathbb{C}^N$ can be identified with the quasi-projective variety

$$\{(x_0 : ... : x_n) \in \mathbb{CP}^N \big| x_0 \neq 0\},$$

via the mapping $(x_1, ..., x_N) \rightarrow (1 : x_1 : ... : x_N)$ with inverse

$$(x_0 : ... : x_N) \rightarrow \left(\frac{x_1}{x_0}, ..., \frac{x_N}{x_0} \right).$$

Using the same mapping any affine variety in $\mathbb{C}^N$ defined by the vanishing of polynomials $f_1, ..., f_m$ of degrees $d_1, ..., d_m$ in N variables is identified with the quasi-projective variety

$$\{(x_0 : ... : x_N) \in \mathbb{CP}^N \big| x_0 \neq 0, \quad \hat{f}_j(x_0, ..., x_N) = 0, \quad 1 \leq j \leq m\}$$

where

$$\hat{f}_j(x_0, ..., x_N) = x_0^{d_j} f_j \left(\frac{x_1}{x_0}, ..., \frac{x_N}{x_0} \right).$$

Any quasi-projective variety X is an open subset of its closure in $\mathbb{CP}^N$ which is a projective variety.

A point x of X is called **non-singular** if there is an open neighborhood U of x in $\mathbb{CP}^N$ and homogeneous polynomials $f_1, ..., f_m$ in $N+1$ variables such that

$$X \cap U = \{(x_0 : ... : x_N) \in U \big| f_j(x_0, ..., x_N) = 0, \quad 1 \leq j \leq m\}$$

and the Jacobian matrix of partial derivatives $\frac{\partial f_j}{\partial x_i}$ has rank m at x. Otherwise x is called a **singular point** of X. The set X_{nonsing} of non-singular points of X is a dense open subset of X, and each connected component is a complex submanifold of $\mathbb{CP}^N$. The variety X is said to have **pure dimension** n if each connected component of X_{nonsing} is a complex manifold of complex dimension n. It is said to be **irreducible** if it cannot be expressed as the union of two closed subvarieties Y and Z unless either Y or Z is X itself.

Examples 4.10.1. The variety $X = \{(x : y : z) \in \mathbb{CP}^2 | yz = 0\}$ is not irreducible; topologically it looks like:

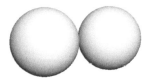

The variety $X = \{(x : y : z) \in \mathbb{CP}^2 | x^3 + y^3 = xyz\}$ is irreducible; topologically it looks like:

Any quasi-projective variety is the union of finitely many irreducible quasi-projective subvarieties $X_1, ..., X_k$ such that X_i is not contained in X_j if $i \neq j$. The subvarieties $X_1, ..., X_k$ are called irreducible components of X. It is easy to check that X has pure dimension n if and only if

$$(X_j)_{\text{nonsing}} = X_j - \{\text{singular points of } X_j\}$$

is a complex manifold of dimension n for each j. A variety of pure dimension one is called a **curve**. A variety of pure dimension two is called a **surface**.

Whitney stratifications

What we need to define intersection homology is a suitable stratification. For quasi-projective varieties we use a Whitney stratification.

Let X be a quasi-projective variety of pure dimension n.

Definition 4.10.2. A **Whitney stratification** of X is given by a filtration

$$X = X_n \supseteq X_{n-1} \supseteq \cdots \supseteq X_0$$

of X by closed subvarieties X_j such that for each j the locally closed subvariety

$$X_j - X_{j-1}$$

is either empty or is a **non-singular** quasi-projective variety of pure dimension j. The connected components S_α of the subvarieties $X_j - X_{j-1}$ are called the **strata** of the stratification and are required to satisfy Whitney's conditions (a) and (b) from [179].

- **Whitney's condition (a)** If a sequence of points $a_i \in S_\alpha$ tends to a point $c \in S_\beta$ then the tangent space $T_c S_\beta$ of c at S_β is contained in the limit of the tangent spaces $T_{a_i} S_\alpha$, provided that this limit exists.

- **Whitney's condition (b)** If a sequence of points $b_i \in S_\beta$ and $a_i \in S_\alpha$ both tend to the same point $c \in S_\beta$ then the limit of the lines joining a_i and b_i is contained in the limit of the tangent spaces to S_α at a_i, provided that both limits exist.

(Here we think of the tangent spaces, and the lines joining pairs of points as linear subspaces of the ambient projective space.)

Roughly speaking, the object of these conditions is to ensure that the normal structure to each stratum S_β is constant along S_β. They imply that for any points x and y on S_β there is a homeomorphism of X to itself which preserves all the strata and takes x to y (this follows from Theorem 4.10.5 below).

Example 4.10.3. Consider the quasi-projective variety

$$X = \{(x, y, z) \in \mathbb{C}^3 \mid x^4 + y^4 = xyz\}.$$

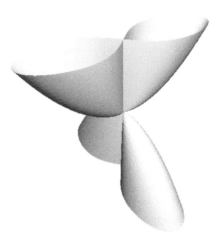

Let $X_2 = X$, let X_1 be the z-axis and let X_0 be empty. This defines a stratification of X with two strata

$$S_\alpha = X - X_1 \text{ and } S_\beta = X_1,$$

both non-singular. This stratification fails Whitney's condition (b). Consider sequences of points a_i to b_i in S_α and S_β chosen as in the diagram below so that the a_i are converging to c much faster than the b_i are. Then the lines joining a_i to b_i will tend to the vertical line through c, while the tangent spaces $T_{a_i} S_\alpha$ tend to the horizontal plane through c.

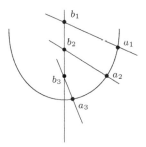

To obtain a Whitney stratification we must take $c \in X_0$. If X_2 and X_1 are defined as before and $X_0 = \{c\}$ then we have a Whitney stratification of X.

Theorem 4.10.4 (Whitney [179, Thm. 19.2]). *Any quasi-projective variety X of pure dimension n has a Whitney stratification.*

We shall define the intersection homology of X using a fixed Whitney stratification.

Theorem 4.10.5 (Borel [24, IV §2]). *Any Whitney stratification*

$$X = X_n \supseteq X_{n-1} \supseteq \cdots \supseteq X_0$$

of a complex quasi-projective variety X of pure dimension n makes X into a topological pseudomanifold of dimension $2n$ with filtration

$$Y_{2n} \supseteq Y_{2n-1} \supseteq \cdots \supseteq Y_0$$

defined by $Y_{2j} = Y_{2j+1} = X_j$.

Furthermore if we want to work simplicially then the following theorem tells us we are free to do so.

Theorem 4.10.6 (Lojasiewicz [114], [115], Goresky [66]). *Let*

$$X = X_n \supseteq X_{n-1} \supseteq \cdots \supseteq X_0$$

be a Whitney stratification of a complex quasi-projective variety X of pure dimension n. Then there is a triangulation of X compatible with the stratification.

Normalisation and curves

We can use a normalisation to compute the intersection homology of a curve (a complex quasi-projective variety of pure dimension 1).

Definition 4.10.7. A quasi-projective complex variety X is called **normal** if the stalk at x of the sheaf of regular functions on X is an integrally closed ring for every $x \in X$. It can be shown using Zariski's Main Theorem (Hartshorne [79, Ch. V Thm. 5.2]) that if a quasi-projective complex variety X is normal in the algebraic sense then it is topologically normal.

Any quasi-projective variety X has a **normalisation** $\pi \colon \tilde{X} \to X$. Here $\tilde{X}$ is a normal quasi-projective variety and π is a finite-to-one surjective-holomorphic map (with a suitable universal property) which restricts to an isomorphism over the non-singular part X_{nonsing} of X.

The normalisation $\tilde{X}$ of a curve X is always non-singular (Hartshorne [79, Ch. III Ex. 5.8]), and hence by Section 4.4 and Proposition 4.5.2 we have

$$IH_i(X) \cong H_i(\tilde{X}). \tag{4.11}$$

In general this does not hold for higher-dimensional varieties since the normalisation need not be non-singular.

The Kähler package

We noted in Chapter 1 that the intersection homology of a complex projective variety satisfies a set of theorems collectively termed the Kähler package. These are

1. **Poincaré duality.** This was proved using simplicial techniques in Goresky and MacPherson's original paper [68] and using sheaf theory in [70]. We will discuss Poincaré duality for intersection homology in Chapter 5 and sketch the sheaf-theoretic proof in Chapter 7.

2. **Lefschetz hyperplane theorem.** Proofs of the Lefschetz hyperplane theorem using stratified Morse theory and sheaf theory (following an idea of Deligne's) can be found in Goresky and MacPherson [71] and [70] respectively. Both proofs hold for a wider range of perversities than the middle. To be precise, let X be an n-dimensional complex projective variety, H a hyperplane which is transverse to the strata of some Whitney stratification of X, and p a perversity for which $p(c) \leq c$ for all c. Then the map

$$I^p H_i(X \cap H) \to I^p H_i(X)$$

is an isomorphism for $i < n - 1$ and a surjection for $i = n - 1$. Goresky and MacPherson [70, §8] point out several interesting consequences for the homology and cohomology of a *normal* complex projective variety X.

(a) Take p to be the zero perversity. Using Proposition 4.5.1 we can deduce that the Gysin map

$$H^i(X \cap H) \to H^{i+2}(X)$$

is an isomorphism for $i > n - 1$ and surjective for $i = n - 1$.

(b) We can show that $IH_1(X) \cong H_1(X)$. By repeatedly applying the Lefschetz hyperplane theorem we deduce that $IH_1(X)$ is isomorphic to the first intersection homology of a surface Y with isolated singularities. Direct computation shows that this group, and hence $H_1(X)$, has even dimension.

(c) Suppose further that X is a local complete intersection. Then the natural maps

$$H_i(X) \to IH_i(X)$$

are isomorphisms (Goresky and MacPherson [70, Prop. 5.6.3]). Hence the homology groups of a normal local complete intersection satisfy the Lefschetz hyperplane theorem.

3. **Hard Lefschetz theorem.** This is a consequence of Beilinson, Bernstein, Deligne and Gabber's decomposition theorem which was proved in [13]. A different proof, using mixed Hodge modules, can be found in Saito [151]. We discuss the decomposition theorem in §8.4.

The hard Lefschetz theorem has many important consequences; here is a simple example. Suppose X is an n-dimensional complex projective variety in $\mathbb{CP}^N$ and that Y is the complex cone on X, i.e. Y is the affine variety in $\mathbb{C}^{N+1}$ cut-out by the homogeneous polynomials in $N + 1$ variables which define X. Let E be the tautological line bundle on X whose fibre over $x \in X$ is the line in $\mathbb{C}^{N+1}$ represented by the point $x \in \mathbb{CP}^N$. (From a different point of view E is the blow-up of Y at the origin and the zero section $X \subset E$ is the exceptional divisor.)

The vertex $\{0\}$ of the complex cone is a singularity of real codimension $2(n + 1)$ and so

$$IH^i(Y) = \begin{cases} IH^i(Y - \{0\}) & i \leq n \\ 0 & \text{otherwise.} \end{cases}$$

The hard Lefschetz theorem states that multiplication by the Euler class of E induces a map

$$L : IH^i(X) \to IH^{i+2}(X)$$

which is injective for $i < n$, surjective for $i + 2 > n$ and such that

$$L^i : IH^{n-i}(X) \to IH^{n+i}(X) \tag{4.12}$$

is an isomorphism for each $i \geq 0$. It follows that the Gysin sequence for E breaks up into short exact sequences

$$0 \to IH^{i-2}(X) \to IH^i(X) \to IH^i(E - X) \to 0$$

for $i \leq n$. We define the primitive part of the intersection cohomology to consist of those classes which are not in the image of L, i.e. not multiples of the Euler class. From (4.12) we have

$$IH^{n-i}_{\mathrm{prim}}(X) = \begin{cases} \ker\left(L^{i+1} : IH^{n-i}(X) \to IH^{n+i}(X)\right) & i \geq 0 \\ 0 & i < 0. \end{cases}$$

Noting that $Y - \{0\}$ and $E - X$ are naturally isomorphic we see that

$$IH^i(Y) \cong IH^i(X - E) \cong IH^i(X)/\mathrm{im}\, L \cong IH^i_{\mathrm{prim}}(X)$$

for $i \leq n$ and trivially $IH^i(Y) = 0 = IH^i_{\mathrm{prim}}(X)$ for $i > n$. Hence we can identify $IH^*(Y)$ with the primitive part of $IH^i(X)$. This can be used in some cases to compute intersection cohomology as a subspace of the cohomology of a non-singular resolution obtained by successive blow-ups — see, for example, Kirwan [111].

4. **Hodge decomposition and Hodge signature theorem.** The Hodge-theoretic parts of the Kähler package are proved in Saito [151] (see Saito [150] for a survey or Looijenga [117] for an introduction).

4.11   Further reading

There are many notions of 'stratified space'. Hughes and Weinberger [82] is a nice survey of some of these from the perspective of topology. Pflaum [142] gives a rather different view from the standpoint of differential geometry. MacPherson's notes [125] contain an appendix with a lovely explanation of 'what a singular space should be' (amongst many other things).

The full details of the simplicial and singular definitions of intersection homology can be found in Goresky and MacPherson [68] and King [110] respectively. The latter also contains a proof of the topological invariance of the intersection homology groups. (We will sketch a different proof using sheaf-theoretic methods, following Goresky and MacPherson [70], in the next chapter.) Another account of the simplicial theory can be found in the book [24] by Borel *et al.* which also includes some interesting examples, a thought-provoking list of problems and conjectures, and a detailed treatment of the sheaf-theoretic approach to intersection homology (which is the subject of the next chapter).

A slightly different approach to intersection homology, using geometric chains, is explained in Goresky and MacPherson [71] and treated more fully in MacPherson [125]. These references explain the close connection between stratified Morse theory and intersection homology (see also Goresky and MacPherson [74]).

Rourke and Sanderson [148] give yet another definition of the intersection homology groups of a PL-space X, essentially as the bordism groups of pseudomanifolds equipped with allowable (in a sense similar to that for singular intersection homology) maps to X. From their point of view intersection homology appears as the stratification-independent theory in a wider family of 'permutation homology' groups associated to a stratified space. They also prove the topological invariance of intersection homology (for PL-spaces) starting from this definition.

Chapter 5

Witt spaces and duality

In this chapter we will take the field $\mathbb{F}$ of coefficients to be the rationals $\mathbb{Q}$.

5.1 Generalised Poincaré duality

If M is a compact oriented topological d-manifold then Poincaré duality states that there is an isomorphism

$$H_i(M) \to \mathrm{Hom}(H_{d-i}(M), \mathbb{Q}) = H_{d-i}(M)^\vee \tag{5.1}$$

between the ith homology (with coefficients in the field $\mathbb{Q}$) and the dual of the $(d - i)$th homology. This is one of the most important properties of the homology of a manifold.

Another way of expressing this is to say that there is a non-degenerate bilinear form, the intersection form,

$$\cap : H_i(M) \times H_{d-i}(M) \to \mathbb{Q}.$$

If we work with simplicial homology groups (assuming that M is triangulable) then this form arises geometrically as follows: any $a \in H_i(M)$ and $b \in H_{d-i}(M)$ can be represented by simplicial chains $\xi \in C_i(M)$ and $\eta \in C_{d-i}(M)$ such that the supports $|\xi|$ and $|\eta|$ meet only in finitely many points. The number of these points, counted with appropriate weights depending on the orientation and coefficients of the chains ξ and η, is an element of $\mathbb{Q}$ which is independent of the choice of ξ and η and is denoted $a \cap b$. Moreover, if $a \neq 0$ there exists some b such that $a \cap b \neq 0$.

If M is not compact then we have a similar result provided we replace $H_{d-i}(M)$ by homology with closed supports $H_{d-i}^{cl}(M)$. In other words there is an isomorphism

$$H_i(M) \to H_{d-i}^{cl}(M)^\vee$$

inducing a non-degenerate bilinear form

$$\cap : H_i(M) \times H_{d-i}^{cl}(M) \to \mathbb{Q},$$

and this can be interpreted geometrically as being given by the intersection number of a compact and a closed support chain. This duality for non-compact manifolds is sometimes called Borel–Moore duality but we will refer to it simply as Poincaré duality. As an example take M to be $\mathbb{R}^m$ then we have

$$H_i(\mathbb{R}^m) = \left\{ \begin{array}{ll} \mathbb{Q} & i = 0 \\ 0 & i \neq 0 \end{array} \right. \quad \text{and} \quad H_i^{cl}(\mathbb{R}^m) = \left\{ \begin{array}{ll} \mathbb{Q} & i = m \\ 0 & i \neq m. \end{array} \right. \tag{5.2}$$

An orientation of $\mathbb{R}^m$ induces an isomorphism

$$H_0(\mathbb{R}^m) \cong H_m^{cl}(\mathbb{R}^m)^\vee$$

of the non-trivial groups. Indeed, when M is compact then we can deduce (5.1) from this non-compact Poincaré duality by finding a 'good' cover of M by finitely many neighbourhoods homeomorphic to $\mathbb{R}^m$ and applying the Mayer–Vietoris sequence repeatedly — see Bott and Tu [26, I §5]. With a little more homological algebra this argument can be extended to Poincaré duality for non-compact manifolds too.

It is easy to see that Poincaré duality does not hold for most singular spaces. One of the primary motivations in the development of intersection homology was to obtain groups which satisfied a suitably generalised version of Poincaré duality.

Theorem 5.1.1 (Generalised Poincaré Duality, Goresky and MacPherson [70, §5.3] cf. [68, §3.3]). *Suppose that X is an oriented topological pseudomanifold of dimension d. Then if p and q are complementary perversities there is a non-degenerate bilinear form*

$$I^p H_i(X) \times I^q H_{d-i}^{cl}(X) \to \mathbb{Q}.$$

This form can be interpreted geometrically. Fix a stratification $X = X_d \supseteq X_{d-2} \supseteq \cdots \supseteq X_0$ of X and, for simplicity, assume X has a compatible triangulation. We work with simplicial intersection chains defined with respect to this stratification. Any $a \in I^p H_i(X)$ and $b \in I^q H_{d-i}^{cl}(X)$ can be represented by $\xi \in I^p C_i(X)$ and $\eta \in I^q C_{d-i}^{cl}(X)$ such that the supports $|\xi|$ and $|\eta|$ meet only in the non-singular part $X - X_{d-2}$ and they meet in finitely many points. The number of these points counted with appropriate weights depending on the orientation and coefficients of the chains ξ and η is an element of $\mathbb{Q}$ which is independent of the choice of ξ and η and is denoted $a \cap b$. Moreover if $a \neq 0$ there exists some b such that $a \cap b \neq 0$.

To understand why Theorem 5.1.1 is true we recall the computation of the intersection homology of a cone from §4.7 (using the description in terms of closed support rather than relative groups). Suppose L is a compact topological m-pseudomanifold and $C(L)$ is the open cone on L. Then

$$I^p H_i\big(C(L)\big) \cong \left\{ \begin{array}{ll} I^p H_i(L) & i < d - p(d+1) \\ 0 & \text{otherwise} \end{array} \right.$$

and

$$I^p H_i^{cl}\big(C(L)\big) \cong \begin{cases} 0 & i \leq d - p(d+1) \\ I^p H_{i-1}(L) & \text{otherwise,} \end{cases}$$

and similarly for the complementary perversity q. If we suppose that the intersection homology of L satisfies generalised Poincaré duality then we have isomorphisms

$$I^p H_i(L) \cong I^q H_{d-i}(L)^\vee.$$

Combining this with the cone calculations for perversities p and q we have

$$\begin{aligned} I^p H_i\big(C(L)\big) &\cong I^p H_i(L) \\ &\cong I^q H_{d-i}(L)^\vee \\ &\cong I^q H_{m+1-i}^{cl}\big(C(L)\big)^\vee \end{aligned}$$

for $i < d - p(d+1)$, or equivalently $d - i \geq d - q(d+1)$, and

$$I^p H_i\big(C(L)\big) = 0 = I^q H_{m+1-i}\big(C(L)\big)$$

for $i \geq d - p(d+1)$, or equivalently, $d - i < d - q(d+1)$. This indicates that

Generalised Poincaré duality for $L \Rightarrow$ Generalised Poincaré duality for $C(L)$

(with a little more care we can check that the induced isomorphisms do indeed come from the intersection form). Thus Theorem 5.1.1 could be proved by choosing a 'good' cover of X by open subsets homeomorphic to cones on lower dimensional pseudomanifolds, deducing by induction on dimension (using the above calculation) that we have (non-compact) Poincaré duality for each of these open subsets and then using Mayer–Vietoris sequences to patch. However, this would be rather fiddly. Instead a more elegant proof involving sheaf theory (which is designed for this sort of local-to-global argument) will be sketched in §7.4. An elementary proof in the simplicial setting can be found in Goresky and MacPherson [68, §3.3].

5.2 Witt spaces

The generalised Poincaré duality of Theorem 5.1.1 is useful but it relates intersection homology groups defined with respect to complementary perversities. In general there is no perversity which is self-complementary so there is no perversity p such that the groups $I^p H_*(X)$ are dual to one another. However, if we restrict X to be a Witt space then the lower middle perversity intersection homology groups have this property.

Definition 5.2.1 (Siegel [162, Prop. 2.5]). A topological pseudomanifold X of dimension d with a given stratification $X = X_d \supseteq X_{d-2} \supseteq \cdots \supseteq X_0$ is a **stratified Witt space** if for each stratum S of odd codimension $2r + 1$ the (lower middle perversity) intersection homology of the link L_S satisfies

$$IH_r(L_S) = 0.$$

There is no condition on strata of even codimension.

It turns out that if X is a Witt space when it is given some stratification then it is a Witt space when given *any* stratification (see Goresky and MacPherson [70, §5.6.1] and also Siegel [162, Defn. 2.1] for piecewise-linear pseudomanifolds). We will say X is a **Witt space** if it is homeomorphic to a stratified Witt space.

Examples 5.2.2.   1. Manifolds are Witt spaces.

2. Any pseudomanifold which can be stratified with only even codimension strata is a Witt space. In particular, any complex quasi-projective variety is a Witt space.

3. The suspension of a torus is *not* a Witt space; the links of the suspension points are the original torus which has non-vanishing middle homology.

If p and q are two perversities for which $p(i) \leq q(i)$ for all i then it is easy to see that $I^p S_*(X)$ is a subcomplex of $I^q S_*(X)$ and that for each i there is an induced map

$$I^p H_i(X) \to I^q H_i(X).$$

In particular for the lower middle perversity m and upper middle perversity n there is a natural map

$$I^m H_i(X) \to I^n H_i(X) \tag{5.3}$$

for $0 \leq i \leq \dim X$.

Proposition 5.2.3 (Siegel [162, Thm. 3.4]). *If X is a Witt space then the induced map $I^m H_i(X) \to I^n H_i(X)$ in (5.3) is an isomorphism.*

This is proved for piecewise-linear Witt spaces in Siegel [162]. A sheaf-theoretic proof for topological Witt spaces, using obstruction sequences for relating intersection homology with respect to different perversities, can be found in Goresky and MacPherson [70, §5.5 and §5.6].

Corollary 5.2.4 (Poincaré duality, Siegel [162, Thm. 3.4]). *The (lower middle perversity) intersection homology of an oriented Witt space X satisfies Poincaré duality. In other words there are isomorphisms*

$$IH_i(X) \cong IH^{cl}_{d-i}(X)^\vee$$

for $0 \leq i \leq d = \dim X$, which induce non-degenerate bilinear forms

$$\cap : IH_i(X) \times IH^{cl}_{d-i}(X) \to \mathbb{Q}.$$

This follows immediately from generalised Poincaré duality and the observation that the upper and lower middle perversities are complementary.

Examples 5.2.5. 1. Let X be a $2k$-dimensional compact oriented Witt space with $IH_k(X) = 0$. Then the suspension $\mathrm{Susp}(X)$ is also a Witt space. It is apparent from Exercise 4.7.3 that Poincaré duality for the intersection homology of the suspension follows from Poincaré duality for the intersection homology of X. The reader is encouraged to think about the relation between the intersection pairings for X and its suspension in terms of cycles.

2. Suppose $(M, \partial M)$ is an even-dimensional compact oriented manifold with boundary. Then the space $\widehat{M}$ obtained by coning off the boundary of M is a Witt space. Poincaré duality for the intersection homology of $\widehat{M}$ follows from the calculation in Remark 4.4.2 and Lefschetz duality for a manifold with boundary.

3. Suppose M is an m-dimensional manifold and $E \to M$ an m-dimensional real vector bundle on M. The Thom space $\mathrm{Th}(E)$ of E is the one point compactification of E. It is naturally stratified by E and the compactification point; the link of this point is the sphere bundle $S(E)$. Since $\dim E = 2m$ the Thom space is a Witt space. By Proposition 4.4.1, excision and stratified homotopy invariance we have

$$IH_i\left(\mathrm{Th}(E)\right) \cong \begin{cases} H_i(M) & i < m \\ \mathrm{im}\ (H_i(D(E)) \to H_i(D(E), S(E))) & i = m \\ H_i(D(E), S(E)) & i > m \end{cases}$$

where $D(E)$ is the disk bundle. Cap product with the Thom class τ of E induces the Thom isomorphism

$$H_i(D(E), S(E)) \cong H_{i-m}(M) : \alpha \mapsto \alpha \cap \tau.$$

We deduce from the geometrical description that the pairing on the intersection homology of the Thom space is given in terms of the pairing on the intersection homology of M by

$$\langle \alpha, \beta \rangle_{\mathrm{Th}(E)} = \langle \alpha, \beta \cap \tau \rangle_M = \langle \alpha \cap \tau, \beta \rangle_M.$$

There is a commutative diagram

$$\begin{array}{ccc} H_m(D(E)) & \longrightarrow & H_m(D(E), S(E)) \\ \| & & \downarrow{\scriptstyle \cap\ \tau} \\ H_m(M) & \underset{\cap\ \eta}{\longrightarrow} & H_0(M) \end{array}$$

where the Euler class $\eta = \tau|_M$. It follows that the middle group

$$IH_m(\mathrm{Th}(E)) \cong H_m(D(E), S(E)) \cong H_0(M)$$

is 1-dimensional and is generated by the class of M. The pairing of this class with itself is

$$\langle [M], [M] \rangle_{\text{Th}(E)} = \langle [M], [M] \cap \eta \rangle = e(E)$$

where $e(E)$ is the Euler number.

Remark 5.2.6 (Goresky and MacPherson [68, §6.3]). The calculations in the third example above are still valid for intersection homology with coefficients in $\mathbb{Z}$ (defined in the obvious way). We note that if $e(E) \neq \pm 1$ then the intersection pairing on $IH_*(\text{Th}(E); \mathbb{Z})$ is not unimodular. In other words Poincaré duality with integral coefficients does not generalise to the intersection homology of Witt spaces. There is a more restrictive class of spaces, the intersection homology Poincaré spaces, obtained by imposing an extra condition on the links of even-codimensional strata whose integral intersection homology satisfies Poincaré duality — see Pardon [141].

We can also compute the intersection homology of $\text{Th}(E)$ with coefficients in $\mathbb{Z}/n\mathbb{Z}$. If $n | e(E)$ then $IH_m(\text{Th}(E)) = 0$ and we see that the universal coefficient theorem fails for intersection homology.

5.3   Signatures of Witt spaces

Let X be a compact oriented Witt space of dimension $d = 4r$. By Corollary 5.2.4 there is a non-degenerate bilinear form, the intersection form,

$$\cap : IH_{2r}(X) \times IH_{2r}(X) \to \mathbb{Q} \tag{5.4}$$

which we can think of geometrically as being given by counting the intersection points of suitable chains representing the classes — see Goresky and MacPherson [68]. Just as the intersection form on the middle homology group of a $4d$-dimensional manifold is *symmetric* so too is this intersection form. It thus has a well-defined signature which is the number of positive eigenvalues minus the number of negative eigenvalues of any matrix representing the form. We will denote the signature of the intersection form (5.4) by $\sigma(X)$. If X is an oriented Witt space of dimension $\not\equiv 0$ modulo 4 then we put $\sigma(X) = 0$. By definition, the signature is a topological invariant of X.

If X is a manifold this definition of signature agrees with the usual one (for which see e.g. Davis and Kirk [51, 3.28]). Signature is a bordism invariant of manifolds; if manifolds M and M' jointly form the boundary of a manifold of one dimension higher then $\sigma(M) = \sigma(M')$ (Davis and Kirk [51, p74]). There is a powerful generalisation of this to the signatures of Witt spaces.

Definition 5.3.1. A **stratified Witt space with boundary** $(X, \partial X)$ of dimension d is a pair of topologically stratified spaces $\partial X \subset X$ such that the inclusion is a stratified map and

1. $X - \partial X$ and ∂X are Witt spaces of respective dimensions d and $d-1$, and

2. ∂X has a neighbourhood in X homeomorphic to the collar $\partial X \times [0,1)$ with ∂X corresponding to the subspace $\partial X \times \{0\}$.

A pair $(X, \partial X)$ of spaces is a **Witt space with boundary** if it is homeomorphic to a stratified Witt space with boundary.

If $X - \partial X$ is oriented then the collaring homeomorphism induces an orientation of the boundary ∂X and we say $(X, \partial X)$ is an **oriented Witt space with boundary**.

We say oriented Witt spaces X and X' are **Witt-bordant** if there is an oriented Witt space with boundary $(Y, \partial Y)$ and an orientation-preserving homeomorphism

$$\partial Y \cong X' \sqcup \overline{X}$$

between the boundary ∂Y and the disjoint union of X' and X with the reversed orientation.

Theorem 5.3.2 (Siegel [162, Thm. 2.1] cf. Goresky and MacPherson [68, §5.2]). *If X and X' are compact oriented Witt spaces which are Witt-bordant then*

$$\sigma(X) = \sigma(X').$$

This certainly implies the bordism invariance of the signature of manifolds but it is much stronger. For instance Siegel [162, Chapter II §3] uses it to give a very simple and geometric proof of Novikov additivity, the result that for manifolds with boundary $(M, \partial M)$ and $(M', \partial M')$ where ∂M is connected and homeomorphic to $\partial M'$ we have

$$\sigma(M \cup_\partial M') = \sigma(M, \partial M) + \sigma(M', \partial M').$$

Here $M \cup_\partial M'$ is the union of M and M' along their boundaries and the signature $\sigma(N, \partial N)$ of a $4r$-manifold with boundary is the signature of the pairing

$$H_{2r}(N, \partial N) \otimes H_{2r}(N) \to \mathbb{Q}$$

arising from Lefschetz duality. This is a good example of how intersection homology not only allows us to understand the topology of singular spaces but also to prove new results, or give simpler proofs of known results, about the topology of manifolds.

5.4 The Witt-bordism groups

In this section we will assume that our Witt spaces are *piecewise-linear*, i.e. they are each equipped with a collection of triangulations such that any refinement is also in the collection and any two triangulations have a common refinement.

The relation of Witt-bordism defines an equivalence relation on oriented Witt spaces (with empty boundary). We define the **Witt-bordism group**

$$\Omega_i^{\text{Witt}}$$

to be the set of equivalence classes of i-dimensional Witt spaces made into an additive group by the operation of disjoint union:

$$[X] + [X'] = [X \sqcup X'].$$

(The empty set is considered to be a Witt space of any dimension and its equivalence class forms the zero. The additive inverse of X is $\overline{X}$, the Witt space X with reversed orientation, since

$$[X] + [\overline{X}] = [X \sqcup \overline{X}] = [\partial (X \times [0,1])] = 0.)$$

Somewhat surprisingly it is possible to give an explicit description of the Witt-bordism groups.

Theorem 5.4.1 (Siegel [162, Prop. 1.1]). *The Witt-bordism groups are given by*

$$\Omega_i^{\text{Witt}} \cong \begin{cases} \mathbb{Z} & i = 0 \\ W(\mathbb{Q}) & i = 4r,\ r > 0 \\ 0 & \text{otherwise,} \end{cases}$$

where $W(\mathbb{Q})$ is the rational Witt group. (The appearance of the rational Witt group is the reason for the name 'Witt space'.)

The rational Witt group $W(\mathbb{Q})$ is a classical algebraic invariant, defined as follows. A **non-degenerate symmetric bilinear form** on a vector space V over $\mathbb{Q}$ is a bilinear map

$$\beta : V \times V \to \mathbb{Q}$$

such that $\beta(v, w) = \beta(w, v)$ and if $v \neq 0$ then there is some w such that $\beta(v, w) \neq 0$. In terms of a basis the form β is represented by a symmetric matrix B with non-zero determinant such that

$$\beta(v, w) = v^t B w.$$

Two forms (not necessarily on the same vector space) are said to be **isometric** if they can be represented by the same matrix. Given forms β and β' on vector spaces V and V' we can form their direct sum $\beta \oplus \beta'$ which is a non-degenerate symmetric bilinear form on $V \oplus V'$. If β and β' are represented by matrices B and B' respectively then the direct sum is represented by the matrix

$$\begin{pmatrix} B & 0 \\ 0 & B' \end{pmatrix}.$$

We say a form β is **hyperbolic** if there is a basis in which it is represented by a matrix

$$\begin{pmatrix} 0 & I_n \\ I_n & 0 \end{pmatrix}$$

where I_n is the $n \times n$ identity matrix. Forms β and β' are said to be **Witt-equivalent** if there exist hyperbolic forms η and η' such that $\beta \oplus \eta$ and $\beta' \oplus \eta'$ are isometric. This is an equivalence relation (prove it!) and the rational Witt group is the set of equivalence classes of non-degenerate symmetric bilinear forms under it. This becomes an additive group under direct sum. In fact, it turns out that a form represents zero in the rational Witt group if and only if it is hyperbolic.

Remark 5.4.2. The structure of the rational Witt group is well-known; see, for example, Milnor and Husemoller [136]. It is

$$W(\mathbb{Q}) \cong \mathbb{Z} \oplus \bigoplus_{\text{primes } p} W(\mathbb{F}_p)$$

where

$$W(\mathbb{F}_p) \cong \begin{cases} \mathbb{Z}/2 & p = 2 \\ \mathbb{Z}/2 \oplus \mathbb{Z}/2 & p = 1 \bmod 4 \\ \mathbb{Z}/4 & p = 3 \bmod 4. \end{cases}$$

The inital $\mathbb{Z}$ corresponds to the signature of the form, i.e. the number of positive eigenvalues minus the number of negative eigenvalues of any representing matrix.

Sketch proof of Theorem 5.4.1. Note that a compact oriented Witt space of dimension 0 is simply a finite collection of oriented points. It easily follows that $\Omega_0^{\text{Witt}} \cong \mathbb{Z}$.

If X is an odd-dimensional compact oriented Witt space then the closed cone $X \times [0,1]/X \times \{0\}$ is a Witt space with boundary X and so $[X] = 0$ in the Witt-bordism group. (The only thing to check here is that the closed cone is a Witt space and the only possible problem is that the Witt space condition might fail at the vertex of the cone. However, since this is a stratum with *even* codimension the condition is vacuous.)

The most interesting case is when the dimension of X is a multiple of 4, say $4r$. Then the intersection form

$$\cap : IH_{2r}(X) \times IH_{2r}(X) \to \mathbb{Q}$$

is symmetric (and non-degenerate by Poincaré duality) and so generates a class, called the Witt class $w(X)$ of X, in $W(\mathbb{Q})$. Siegel shows that $w(X) = w(X')$ for Witt-bordant X and X' using much the same ideas used to prove that the signature is a Witt-bordism invariant. Hence we obtain a map

$$w : \Omega_{4r}^{\text{Witt}} \to W(\mathbb{Q}) \tag{5.5}$$

which we wish to show is an isomorphism.

Suppose $w(X) = 0$. Then the intersection form of X must be hyperbolic, in other words we can choose a basis $\{v_1, \ldots, v_n, w_1, \ldots, w_n\}$ of $IH_{2r}(X)$ with respect to which the intersection form is represented by

$$\begin{pmatrix} 0 & I_n \\ I_n & 0 \end{pmatrix}.$$

In this situation Siegel constructs a compact oriented Witt space $\tilde{X}$ which is Witt-bordant to X and such that there is a short exact sequence

$$0 \to \langle v_1, w_1 \rangle \to IH_{2r}(X) \to IH_{2r}(\tilde{X}) \to 0$$

where $\langle v_1, w_1 \rangle$ is the subspace generated by v_1 and w_1. Roughly $\tilde{X}$ is formed by collapsing a simplicial cycle representing the class v_1, a process which Siegel terms **elementary surgery**. By repeating this we can successively kill off pairs $\{v_i, w_i\}$ of elements in $IH_{2r}(X)$ until we obtain a Witt space Y which is Witt-bordant to X and with

$$IH_{2r}(Y) = 0.$$

It follows that Y, and thereby X itself, is Witt-bordant to the empty set because the closed cone on Y is a Witt space with boundary Y. So the map w in (5.5) is injective.

To show that w is surjective Siegel uses **plumbing**: this is a procedure which, given a symmetric non-degenerate matrix B (satisfying certain conditions) and an integer $r \geq 1$, constructs a compact oriented $4r$-dimensional manifold with boundary $(M_B, \partial M_B)$ whose intersection form

$$H_{2r}(M_B) \times H_{2r}(M_B, \partial M_B) \to \mathbb{Q}$$

is represented by the matrix B in a natural basis — see Browder [35]. By glueing the closed cone on ∂M_B onto the boundary we obtain a compact oriented Witt space whose intersection form

$$IH_{2r}(\widehat{M_B}) \times IH_{2r}(\widehat{M_B}) \to \mathbb{Q}$$

is again represented by B. The surjectivity of w follows because for any class in the Witt group we can find a representative form with appropriate matrix B and then by the above construct a Witt space with precisely that intersection form.

Finally, we need to show that $\Omega_{4r+2}^{\text{Witt}} = 0$, i.e. that all $(4r+2)$-dimensional compact oriented Witt spaces are Witt-bordant to the empty set. The key point is that the intersection form of a $(4r+2)$-dimensional Witt space is anti-symmetric and so can be represented by a matrix with block form

$$\begin{pmatrix} 0 & I_n \\ -I_n & 0 \end{pmatrix}.$$

Thus we can construct an explicit Witt-bordism to the empty set using Siegel's surgery procedure in almost exactly the same way that we showed w was injective above. □

5.5   Further reading

In addition to a signature, a Whitney stratified (in the sense of Goresky and MacPherson [74, §1.2]) Witt space possesses an invariant called the **L-class**. When X is a smooth manifold then the L-class is Poincaré dual to the Hirzebruch L-class (see Milnor and Stasheff [137] for the definition). More generally, the L-class $L(X) \in H_*(X; \mathbb{Q})$ of a Whitney stratified Witt space X is a class in (ordinary) rational homology which contains information about the signature of certain subspaces of X. To be precise, a subspace Y of X is **normally non-singular of codimension k with trivial normal bundle** if there is a neighbourhood U of Y in X and a homeomorphism

$$Y \times \mathbb{R}^k \to U$$

which is the identity on Y. Any normally non-singular subspace Y determines a cohomology class $[Y]^* \in H^k(X; \mathbb{Q})$ (Cappell and Shaneson [40, §5]) and the L-class has the property that if Y has trivial normal bundle then

$$[Y]^* (L(X)) = \sigma(Y).$$

In particular, if we suppose X is connected so that $H_0(X; \mathbb{Q}) \cong \mathbb{Q}$ then the piece of $L(X)$ in $H_0(X; \mathbb{Q})$ is simply the signature $\sigma(X)$ of X.

The L-class was initially constructed geometrically for spaces with only even codimensional strata in Goresky and MacPherson [68, §5] although identical methods can be used to construct it for arbitrary Witt spaces (Banagl, Cappell and Shaneson [6, §2]). A more abstract, sheaf-theoretic construction can be found in Cappell and Shaneson [40, §5].

Siegel's result on the Witt bordism groups has several important consequences, most of which arise from the fact that the Witt bordism groups form a generalised homology theory (this is a property of bordism rather than a special property of Witt spaces). Siegel's calculation identifies the groups of a point and it is not difficult to use this to prove that Witt bordism gives a homology theory called rational L-theory which arises in the classification of manifolds. Hughes and Weinberger [82, §2] contains a nice discussion.

Furthermore, there is an analogous result for integral rather than rational coefficients, see Pardon [141]. That is, there is a class of piecewise-linear pseudomanifolds, dubbed **intersection homology Poincaré spaces**, whose intersection homology groups with coefficients in $\mathbb{Z}$ obey Poincaré duality and whose bordism groups are given by an appropriate generalisation of the rational Witt group. To be precise they are isomorphic to the symmetric L-groups $L^*(\mathbb{Z})$ of Mischenko and Ranicki (see Ranicki [144] for the definition).

The bordism of Witt spaces provides a geometric generalisation of the rational Witt group. Parallel (orthogonal?) to this there are algebraic generalisations involving bordism groups of self-dual complexes of sheaves. Cappell and Shaneson [40] use these to prove a beautiful result relating the L-classes of stratified spaces with a stratified map between them (see also Youssin [180] and Banagl, Cappell and Shaneson [6]). This result is a broad generalisation of classical results about the L-classes of fibre bundles. It can also be seen as an analogue of the decomposition theorem (see Remark 8.4.4).

Chapter 6

L^2-cohomology and intersection cohomology

In this chapter we will briefly discuss Cheeger's L^2-cohomology and its relation to intersection cohomology. In particular we will discuss two classes of examples, varieties with locally conical singularities and locally symmetric varieties, for which the two theories are isomorphic. We will work with coefficients in $\mathbb{R}$, except in §6.4 where we take coefficients in $\mathbb{C}$.

6.1 L^2-cohomology and Hodge theory

Suppose M is a smooth manifold with a Riemannian metric. In the same way as in §1.3 we can define L^2-cohomology groups of M. In brief, let $A^r(M)$ be the space of differential r-forms on M. The metric induces an inner product on each $A^r(M)$ and we say a form $\omega \in A^r(M)$ is in the subspace $L^r(M)$ if it is square-integrable with respect to the associated norm, i.e. if

$$\int_M \|\omega\|^2 < \infty.$$

The L^2-cohomology group $H^r_{(2)}(M; \mathbb{R})$ of M (with *real* coefficients) is defined to be

$$\frac{\{\omega \in L^r(M) \mid d\omega = 0\}}{\{\eta \in L^r(M) \mid \exists \zeta \in L^{r-1}(M), d\zeta = \eta\}}.$$

When M is compact all forms are square-integrable and this reduces to the usual definition of the de Rham cohomology group. However, if M is not compact then the L^2-cohomology can be significantly different from the de Rham cohomology; indeed it is not even necessarily finite-dimensional.

The L^2-cohomology groups of M depend upon the metric, but only up to quasi-isometry.

Definition 6.1.1. Two Riemannian metrics g and h on a manifold M are called **quasi-isometric** if there exists a positive constant K such that at every point p of M the inner products g_p and h_p on the tangent space T_pM satisfy the inequalities:

$$K^{-1}g_p \leq h_p \leq Kg_p.$$

The norms defined by the metrics g and h then satisfy corresponding inequalities at each point. In particular if ω is a differential r-form on Y then ω is square-integrable with respect to the norm defined by the metric g, if and only if ω is square-integrable with respect to the norm defined by the metric h. Thus the L^2-cohomology groups of M defined using two quasi-isometric metrics are the same.

Hodge theory allows us to show that the de Rham cohomology of a compact oriented manifold satisfies Poincaré duality. We briefly recall the argument, the details can be found in Warner [175, §6]. We choose a metric on our manifold M^m and define the **Hodge star** operator

$$* : A^r(M) \rightarrow A^{m-r}(M)$$

to be the unique operator satisfying

$$\alpha \wedge *\beta = (\alpha, \beta)\Omega$$

for all $\alpha \in A^r(M)$ where the round brackets denote the inner product arising from the metric and Ω is the volume form. The **Laplacian** is the second order differential operator

$$\Delta = d\delta + \delta d$$

where $\delta = *d*$. (In fact δ is the adjoint of d with respect to the inner product but in this infinite-dimensional setting it is easier to define it using the Hodge star.) A differential form ω is said to be **harmonic** if $\Delta\omega = 0$.

On a compact Riemannian manifold the harmonic forms are both closed and coclosed, i.e. $d\omega = 0$ and $\delta\omega = 0$. It follows that each harmonic form determines a cohomology class. Furthermore the Hodge theorem tells us that there is a unique harmonic form in each cohomology class; the de Rham cohomology is isomorphic to the space of harmonic forms. It is clear that if ω is harmonic then so is $*\omega$. Thus the Hodge star furnishes us with an isomorphism

$$H^r(M; \mathbb{R}) \cong H^{m-r}(M; \mathbb{R})$$

which can be identified as the Poincaré duality isomorphism.

It is natural to ask how much of this theory carries over to L^2-cohomology. We can clearly still define the Hodge star, the Laplacian and harmonic forms. However, it is no longer the case that a harmonic form in $L^r(M)$ is necessarily closed and coclosed (for example the function $x^2 - y^2$ on the unit disc equipped with the Euclidean metric is harmonic and square-integrable but it is not closed, i.e. not constant). In particular a harmonic form may no longer define a cohomology class. Therefore we consider the subspace $\mathcal{H}^r(M)$ of closed and coclosed harmonic forms in $L^r(M)$. There is a natural map

$$\mathcal{H}^r(M) \to H^r_{(2)}(M; \mathbb{R}) \tag{6.1}$$

taking a form to its L^2-cohomology class. It is not necessarily an isomorphism but it is known that if the metric on M is *complete* then all harmonic forms are closed and coclosed and that (6.1) is an injection (Cheeger, Goresky and MacPherson [48, §3]).

Let us suppose that (6.1) is an isomorphism for M (sometimes it is said that the **strong Hodge theorem** holds for M). It is easy to see that the Hodge star defines an isomorphism

$$\mathcal{H}^r(M) \cong \mathcal{H}^{m-r}(M)$$

and it is an immediate consequence that the L^2-cohomology of M has Poincaré duality. This suggests that it might be possible to interpret the L^2-cohomology as the intersection cohomology of some singular compactification of M. To make this idea more precise let us suppose that M is the non-singular part of a complex projective variety X, so that X is a natural compactification of M. Let β be a square-integrable differential i-form on

$$M = X - \Sigma$$

such that $d\beta$ is also square-integrable. Then one can show that for almost all intersection chains $\xi \in IC_i(X)$ the integral

$$\int_\xi \beta$$

exists and Stokes' theorem

$$\int_\xi d\beta = \int_{\partial\xi} \beta$$

is satisfied. (Note that the support of an intersection chain $\xi \in IC_i(X)$ is never contained in the singular set Σ where β is not defined: it meets Σ in a subset of dimension at most $i - 2$.) In this way integration can be used to define a natural pairing

$$H^i_{(2)}(X - \Sigma) \otimes IH_i(X) \to \mathbb{R} \tag{6.2}$$

or equivalently a natural map

$$H^i_{(2)}(X - \Sigma) \to (IH_i(X))^\vee = IH^i(X). \tag{6.3}$$

When X is non-singular this map is the de Rham isomorphism

$$H^i_{\mathrm{DR}}(X) \to H^i(X),$$

see Griffiths and Harris [77, p44]. Sections 6.3 and 6.4 will be spent discussing two classes of singular varieties, with metrics on their non-singular parts, for which (6.3) is an isomorphism. In both cases the strong Hodge theorem will also hold and so we will have isomorphisms

$$\mathcal{H}^r(X - \Sigma) \cong H_{(2)}^r(X - \Sigma) \cong IH^*(X).$$

It is conjectured that such isomorphisms hold more generally. For further details see Cheeger [44, 45, 46] and Cheeger, Goresky and MacPherson [48].

In the next section we will give some evidence for this conjecture based on the simplest case, that of a cone on a Riemannian manifold.

6.2 The L^2-cohomology of a punctured cone

Definition 6.2.1. If Y is a compact manifold with Riemannian metric g_Y let $C^*(Y)$ be the **punctured cone**

$$C^*(Y) = C(Y) - \{\text{vertex}\} = (0, 1) \times Y$$

with Riemannian metric

$$g = dt \otimes dt + t^2 \pi^* g_Y$$

where t is the standard coordinate on the open interval $(0, 1)$ and $\pi : (0, 1) \times Y \to Y$ is the projection. (Recall that the cone $C(Y)$ is obtained from the product $[0, 1) \times Y$ by identifying the points of $\{0\} \times Y$ to give a single point which is the vertex of the cone.)

Note that any differential i-form ξ on $C^*(Y)$ can be written uniquely as

$$\xi = \eta + dt \wedge \zeta \tag{6.4}$$

where η and ζ are differential forms which do not involve dt. In other words with respect to (real) local coordinates $(y_1, ..., y_m)$ on Y we can write

$$\eta(t, y) = \sum_{\alpha \in I(i)} \eta_\alpha(t, y) dy^\alpha \tag{6.5}$$

where $I(i)$ is the set of all multi-indices $\alpha = (\alpha_1, ..., \alpha_i)$ such that $1 \leq \alpha_1 < \cdots < \alpha_i \leq m$, where

$$dy^\alpha = dy^{\alpha_1} \wedge \cdots \wedge dy^{\alpha_i}$$

and where η_α is smooth function on $(0, 1) \times Y$. Similarly

$$\zeta(t, y) = \sum_{\alpha \in I(i-1)} \zeta_\alpha(t, y) dy^\alpha. \tag{6.6}$$

Thus for fixed $t \in (0,1)$ we can regard $\eta(t,y)$ and $\zeta(t,y)$ as defining differential forms on Y. The Riemannian metric on $C^*(Y)$ is defined in such a way that the norm of ξ is given by

$$\|\xi(t,y)\|^2 = t^{-2i}\|\eta(t,y)\|_Y^2 + t^{-2(i-1)}\|\zeta(t,y)\|_Y^2 \qquad (6.7)$$

where $\|\ \|_Y$ is the norm induced by the metric g_y on Y. The factor t^{-2i} occurs because $\eta(t,y)$ lies in the ith exterior power of the dual of the tangent space to $C^*(Y)$ at the point (t,y).

Proposition 6.2.2 (Cheeger [45]). *Let Y be a compact Riemannian manifold of dimension m and let $C^*(Y)$ be the punctured cone on Y with the metric defined at 6.2.1. Then*

$$H_{(2)}^i(C^*(Y)) = \begin{cases} H^i(Y) & i \le m/2, \\ 0 & i > m/2. \end{cases}$$

Remark 6.2.3. Note that the L^2-cohomology of Y is the same as its de Rham cohomology (since Y is compact) and hence there is a natural isomorphism

$$H_{(2)}^i(Y) \cong H^i(Y).$$

It is highly suggestive to compare this calculation with that in Proposition 4.7.2 for the (middle perversity) intersection homology of the cone on a pseudomanifold.

Sketch proof of Proposition 6.2.2. Let

$$\pi \colon C^*(Y) = (0,1) \times Y \to Y$$

be the projection. If $\omega \in A^i(Y)$ is a differential i-form on Y then with respect to local coordinates $(y_1, ..., y_m)$ we can write

$$\omega(y) = \sum_{\alpha \in I(i)} \omega_\alpha(y)dy^\alpha .$$

The i-form $\pi^*\omega$ on $C^*(Y)$ is then defined in local coordinates $(t, y_1, ..., y_m)$ by the same formula
$$\pi^*\omega(t,y) = \sum_{\alpha \in I(i)} \omega_\alpha(y)dy^\alpha .$$

By (6.7) we have

$$\|\pi^*\omega(t,y)\|^2 = t^{-2i}\|\omega(y)\|_Y^2.$$

Moreover the volume form on $C^*(Y)$ at a point (t,y) differs from the volume form on Y at y by a factor of t^m so

$$\int_{C^*(Y)} \|\pi^*\omega\|^2 = \int_0^1 \int_Y t^{-2i}\|\omega\|^2 t^m \, dt.$$

Since Y is compact it follows that $\pi^*\omega$ is square-integrable if and only if $\omega = 0$ or

$$\int_0^1 t^{m-2i} dt < \infty.$$

Therefore if $m - 2i > -1$ or equivalently $i \leq \frac{m}{2}$ then π^* restricts to a map

$$\pi^*: L^i(Y) \to L^i(C^*(Y))$$

which commutes with d and hence induces a natural map

$$\pi^*: H^i(Y) \cong H^i_{(2)}(Y) \to H^i_{(2)}(C^*(Y)). \tag{6.8}$$

We shall show that this map is an isomorphism for all $i \leq \frac{m}{2}$. Given a differential i-form ξ on $C^*(Y)$ write

$$\xi = \eta + dt \wedge \zeta$$

as at (6.4). There is an i-form $\partial\eta/\partial t$ on $C^*(Y)$ defined in local coordinates $(y_1, ..., y_m)$ by

$$\frac{\partial\eta}{\partial t}(t, y) = \sum_{\alpha \in I(i)} \frac{\partial\eta_\alpha}{\partial t}(t, y) dy^\alpha$$

in the notation of (6.5). Similarly there is an $(i - 1)$-form $\partial\zeta/\partial t$ given by

$$\frac{\partial\zeta}{\partial t}(t, y) = \sum_{\alpha \in I(i-1)} \frac{\partial\zeta_\alpha}{\partial t}(t, y) dy^\alpha.$$

We can define $d_Y: A^i(C^*(Y)) \to A^{i+1}(C^*(Y))$ in local coordinates $(y_1, ..., y_m)$ by

$$\begin{aligned} d_Y\xi(t, y) &= \sum_{1 \leq j \leq m} \sum_{\alpha \in I(i)} \frac{\partial\eta_\alpha}{\partial y_j}(t, y) dy_j \wedge dy^\alpha \\ &+ \sum_{1 \leq j \leq m} \sum_{\alpha \in I(i-1)} \frac{\partial\zeta_\alpha}{\partial Y_j}(t, y) dy_j \wedge dt \wedge dy^\alpha. \end{aligned}$$

Then

$$d_Y\xi = d_Y\eta - dt \wedge d_Y\zeta,$$

and

$$d\xi = d_Y\xi + dt \wedge \frac{\partial\eta}{\partial t} = d_Y\eta + dt \wedge \left(\frac{\partial\eta}{\partial t} - d_Y\zeta\right).$$

Now we fix $s \in (0, 1)$ and define

$$H: A^i(C^*(Y)) \to A^{i-1}(C^*(Y))$$

in local coordinates $(y_1, ..., y_m)$ by

$$(H\xi)(t, y) = \sum_{\alpha \in I(i-1)} \left(\int_s^t \zeta_\alpha(\tau, y) d\tau \right) dy^\alpha$$

where $\xi = \eta + dt \wedge \zeta$ as before. We shall write this more conveniently as $H\xi = \int_s^t \zeta$. Then

$$
\begin{aligned}
dH\xi &= d_Y \int_s^t \zeta + dt \wedge \frac{\partial}{\partial t} \int_s^t \zeta \\
&= \int_s^t d_Y \zeta + dt \wedge \zeta.
\end{aligned}
$$

Also

$$
\begin{aligned}
Hd\xi &= H \left(d_Y \eta + dt \wedge \left(\frac{\partial \eta}{\partial t} - d_Y \zeta \right) \right) \\
&= \int_s^t \left(\frac{\partial \eta}{\partial t} - d_Y \zeta \right) \\
&= \eta - \pi^* \left(\eta^{(s)} \right) - \int_s^t d_Y \zeta
\end{aligned}
$$

where $\eta^{(s)} \in A^i(Y)$ is given in local coordinates $(t, y_1, ..., y_m)$ by

$$\eta^{(s)}(y) = \sum_{\alpha \in I(i)} \eta_\alpha(s, y) dy^\alpha.$$

Thus

$$
\begin{aligned}
dH\zeta + Hd\zeta &= dt \wedge \zeta + \eta - \pi^* \left(\eta^{(s)} \right) && (6.9) \\
&= \xi - \pi^* \left(\eta^{(s)} \right). && (6.10)
\end{aligned}
$$

Now if ξ is a square-integrable i-form then

$$\int_{C^*(Y)} \|\xi\|^2 = \int_0^1 \int_Y \left(t^{-2i} \|\eta^{(t)}\|_Y^2 + t^{-2(i-1)} \|\zeta^{(t)}\|_Y^2 \right) t^m dt$$

is finite. Since $H\xi$ is an $(i-1)$-form

$$\int_{C^*(Y)} \|H\xi\|^2 \leq \int_0^1 \int_Y t^{-2(i-1)} \int_s^t \|\zeta^{(\tau)} d\tau\|_Y^2 t^m dt.$$

Using the Cauchy–Schwartz inequality and reversing the order of the integration we find if $i \leq \frac{m}{2}$ then

$$\int_{C^*(Y)} \|H\xi\|^2 \leq \int_0^1 \int_Y t^{m-2i+2} \left| \int_s^t \|\zeta^{(\tau)}\|_Y^2 d\tau \right| dt$$

$$= \int_0^s \int_Y \|\zeta^{(\tau)}\|_Y^2 \int_0^\tau t^{m-2i+2} dt d\tau + \int_s^1 \int_Y \|\zeta^{(\tau)}\|_Y^2 \int_\tau^1 t^{m-2i+2} dt d\tau$$

$$\leq \frac{1}{m-2i+3} \left(\int_0^s \int_Y \|\zeta^{(\tau)}\|_Y^2 \tau^{m-2i+3} d\tau + \int_s^1 \int_Y \|\zeta^{(\tau)}\|_Y^2 d\tau \right)$$

$$\leq \left(\frac{1+s^{-m+2i-2}}{m-2i+3} \right) \int_{C^*(Y)} \|\xi\|^2 < \infty.$$

Hence $H\xi$ is square-integrable.

We have shown that if ξ is a square integrable i-form on $C^*(Y)$ and $i \leq \frac{m}{2}$ then $H\xi$ is square-integrable and

$$\xi = dH\xi + Hd\xi + \pi^*(\eta^{(s)}). \tag{6.11}$$

Therefore if $d\xi = 0$ then

$$\xi \in d\left(L^{i-1}(C^*(Y))\right) + \pi^*\left(L^i(Y)\right)$$

so

$$\pi^*: H^i(Y) \to H^i_{(2)}(C^*(Y))$$

is surjective for $i \leq \frac{m}{2}$. Moreover since $d^2 = 0$ we have by (6.11)

$$d\xi = d(Hd\xi) + d\pi^*(\eta^{(s)})$$
$$= d(Hd\xi) + \pi^*(\eta^{(s)}).$$

It comes straight from the definition of H that $Hd\xi = 0$ if $d\xi \in \pi^*(L^i(Y))$, and hence it follows easily that $\pi^*: H^i(Y) \to H^i_{(2)}(C^*(Y))$ is injective for $i \leq \frac{m}{2}$.

It remains to show that $H^i_{(2)}(C^*(Y)) = 0$ for $i > \frac{m}{2}$. The Cauchy–Schwartz inequality tells us that if ϕ is a square-integrable i-form on $C^*(Y)$ and we have $0 < a < b < 1$ then

$$\left(\int_b^a \int_Y \|\phi^{(t)}\|_Y^2 dt \right)^2 \leq \left(\int_0^1 \int_Y t^{m-2i} \|\phi^{(t)}\|_Y^2 dt \right) \left(\int_b^a \int_Y t^{2i-m} dt \right)$$

$$= \left(\int_{C^*(Y)} \|\phi\|^2 \right) \left(\int_Y 1 \right) \left(\frac{b^{2i-m+1} - a^{2i-m+1}}{2i-m+1} \right).$$

Therefore the integral $\int_0^1 \int_Y \|\phi^{(t)}\|_Y dt$ exists if $i \geq \frac{m}{2}$, and so for almost all $y \in Y$ the integral

$$\int_0^t \phi = \int_0^t \phi^{(\tau)} d\tau$$

exists for all $t \in (0, 1)$. The idea is now that if $\xi = \eta + dt \wedge \zeta$ is a square-integrable i-form and $i - 1 \geq \frac{m}{2}$ then we define

$$H^0 \xi = \int_0^t \zeta.$$

The argument used above can be easily modified to show that $H^0 \xi$ is square-integrable and that

$$\xi = dH^0 \xi + H^0 d\xi.$$

In particular if $d\xi = 0$ then $\xi = dH^0 \xi$. From this it can be deduced that $H^i_{(2)}(C^*(Y)) = 0$ when $i - 1 \geq \frac{m}{2}$, though technical difficulties arise because $H^0 \xi$ is not necessarily differentiable.

The only case we have not yet covered is when m is odd and $i = \frac{m+1}{2}$. This case is more delicate but it can be shown that $H^i_{(2)}(C^*(Y)) = 0$ in this case also (see Cheeger [45]). □

6.3 Varieties with isolated conical singularities

When the singularities of the complex projective variety X are particularly simple it is possible to show that

$$H^*_{(2)}(X) \cong IH^*(X)$$

by doing a local calculation for L^2-cohomology and comparing it with the local calculation in §4.7 for intersection cohomology.

Definition 6.3.1. Let $X \subseteq \mathbb{CP}^n$ be a quasi-projective variety with isolated singularities. Let

$$\Sigma = \{x_1, ..., x_q\}$$

be the set of singular points of X. We say that X has **isolated conical singularities** if there exist compact Riemannian manifolds $Y_1, ..., Y_q$ and disjoint open neighbourhoods $U_1, ..., U_q$ of $x_1, ..., x_q$ in X such that U_j is homeomorphic to the cone $C(Y_j)$ and $U_j - \{x_j\}$ is quasi-isometric to the punctured cone $C^*(Y_j)$ for $1 \leq j \leq q$. Here the metric on $U_j - \{x_j\}$ is given by the restriction of the Fubini–Study metric on $\mathbb{CP}^n$ and the metric on $C^*(Y_j)$ is given by the metric defined in Definition 6.2.1.

Suppose that X is a projective variety with isolated conical singularities, and let $\dim_{\mathbb{C}} X = n$.

Lemma 6.3.2. *Every $x \in X$ has arbitrarily small open neighbourhoods U in X such that the natural maps*

$$H^*_{(2)}(U) \to IH^*(U)$$

are isomorphisms.

Proof. It is easy to check that if $x \in X$ is a non-singular point of X then x has arbitrarily small open neighbourhoods U in

$$X_{\text{nonsing}} = X - \Sigma$$

which are quasi-isometric to cones on a sphere. A simpler version of the argument used to prove Proposition 6.2.2 shows that the L^2-cohomology of such a neighbourhood is trivial, i.e.

$$H^i_{(2)}(U) = \begin{cases} \mathbb{R} & i = 0, \\ 0 & \text{otherwise.} \end{cases}$$

On the other hand

$$IH^i(U) \cong H^i(U) \cong \begin{cases} \mathbb{R} & i = 0 \\ 0 & \text{otherwise,} \end{cases}$$

since U is non-singular and contractible. It is clear from the definition of (6.2) that the natural map

$$H^i_{(2)}(U) \to IH^i(U)$$

is non-zero when $i = 0$ and hence is an isomorphism for all $i \geq 0$.

   Now suppose x is a singular point of X. Then since X has isolated conical singularities there is a compact Riemannian manifold Y and an open neighbourhood U of x in X such that U is homeomorphic to the cone $C(Y)$ and $U - \{x\}$ is quasi-isometric to the punctured cone $C^*(Y)$. It is easy to see that U may be chosen arbitrarily small. Then by Proposition 6.2.2 since the real dimension of Y is $2n - 1$ we have

$$H^i_{(2)}(U) \cong \begin{cases} H^i(Y) & i \leq n - 1, \\ 0 & i \geq n. \end{cases} \tag{6.12}$$

On the other hand since U has a single isolated singularity at x it follows from Proposition 4.4.1 that

$$IH_i(U) \cong \begin{cases} H_i(U - \{x\}) & i \leq n - 1, \\ \text{im}\left(H_i(U - \{x\}) \to H_i(U)\right) & i = n, \\ H_i(U) & i \geq n + 1. \end{cases}$$

Moreover since U is contractible we have $H_i(U) = 0$ if $i \geq 1$, and since $U - \{x\}$ is homeomorphic to $C^*(Y) = (0, 1) \times Y$ we have

$$H_i(U - \{x\}) \cong H_i\left((0, 1) \times Y\right) \cong H_i(Y)$$

for all i. Thus,

$$IH_i(U) \cong \begin{cases} H_i(Y) & \text{if } i \leq n - 1, \\ 0 & \text{if } i \geq n. \end{cases} \tag{6.13}$$

Taking duals and comparing with (6.12) we find that $H^i_{(2)}(U) \cong IH^i(U)$ for all i. In order to check that this isomorphism corresponds to the natural map

$$H^i_{(2)}(U) \to IH^i(U)$$

it suffices to consider the case $i \le n-1$. Then the isomorphism $H^i_{(2)}(U) \to H^i(Y)$ of (6.12) is the composition of the inverse of the map

$$\pi^*\colon H^i_{(2)}(Y) \to H^i_{(2)}(U - \{x\}) = H^i_{(2)}(U)$$

induced by the projection

$$\pi\colon U - \{x\} = (0,1) \times Y \to Y$$

with the natural isomorphism $H^i_{(2)}(Y) = H^i_{DR}(Y) \to H^i(Y)$. On the other hand the isomorphism $IH_i(U) \cong H_i(Y)$ of (6.13) is the composition of the identification

$$IH_i(U) = IH_i(U - \{x\}) = H_i(U - \{x\})$$

with the isomorphism

$$\pi_*\colon H_i(U - \{x\}) \to H_i(Y).$$

The result follows. $\qquad\qquad\qquad\qquad\qquad\qquad\qquad\qquad\qquad\qquad$ □

Let X be a projective variety with isolated conical singularities. It turns out that the existence of the natural map

$$H^*_{(2)}(X) \to IH^*(X)$$

together with Lemma 6.3.2 implies the following theorem. (The proof follows fairly easily from the sheaf-theoretic treatment of intersection cohomology we give in the next chapter.)

Theorem 6.3.3 (Cheeger [45]). *The natural map*

$$H^*_{(2)}(X) \to IH^*(X)$$

from the L^2-cohomology of X to its intersection cohomology is an isomorphism.

The strong Hodge theorem for varieties with isolated conical singularities also follows from Lemma 6.3.2, essentially because the L^2-cohomology is then finite-dimensional. The proof can be found in Cheeger, Goresky and MacPherson [48, §3].

6.4   Locally symmetric varieties

Locally symmetric varieties are of great importance in Riemannian geometry, algebraic geometry, number theory and representation theory. There is a vast literature on the area but we will give only the briefest of treatments here, sufficient to state the result (Theorem 6.4.6) in which we are interested. Our treatment is heavily influenced by Zucker's introduction [184]. We will assume some familiarity with the theory of Lie groups, for which see Atiyah *et al.* [4], Carter, Segal and MacDonald [42] and Springer [165].

Let G be a semi-simple Lie group with finite centre and K a maximal compact subgroup. Let $D = G/K$ be the quotient of G by the natural right action of K. Clearly G acts on the left on D making it into a homogeneous space.

There is a unique automorphism θ of G (called the **Cartan involution**) such that

- θ is an involution, i.e. $\theta^2 = \mathrm{id}$,

- K is the fixed point set of θ.

The **Killing form** is the symmetric bilinear form defined by

$$B(x, y) = \mathrm{tr}\left(\mathrm{ad}_{\mathfrak{g}}(x)\mathrm{ad}_{\mathfrak{g}}(y)\right).$$

It is invariant under the adjoint action of K on $\mathfrak{g}$ and restricts to a positive definite form on the negative eigenspace of θ on the Lie algebra $\mathfrak{g}$. This eigenspace is naturally identified with the tangent space to D at the class of the identity. The adjoint action of K preserves this eigenspace and we can define a G-invariant Riemannian metric on D by translation. Furthermore this metric is complete.

Example 6.4.1. Take $G = SP_{2n}(\mathbb{R})$ to be the symplectic group, i.e. the group of invertible linear transformations which preserve the anti-symmetric bilinear form on $\mathbb{R}^{2n}$ represented by the matrix

$$J = \begin{pmatrix} 0 & I_n \\ -I_n & 0 \end{pmatrix}.$$

In concrete terms we have

$$SP_{2n}(\mathbb{R}) = \left\{ X = \begin{pmatrix} A & B \\ -B & A \end{pmatrix} \mid A, B \in M_n(\mathbb{R}), \det X \neq 0 \right\}.$$

The Lie algebra $\mathfrak{sp}_{2n}(\mathbb{R})$ of $SP_{2n}(\mathbb{R})$ can be identified with the same set of matrices where we remove the condition that the determinant is non-zero.

The symplectic group can be naturally identified with a subgroup of the $n \times n$ complex matrices via the injective homomorphism

$$\Phi : \begin{pmatrix} A & B \\ -B & A \end{pmatrix} \mapsto A + iB.$$

Note that $\Phi(JX) = i\Phi(X)$. Under this identification the unitary group U_n corresponds to the maximal compact subgroup K of symplectic matrices with $A^t A + B^t B = 1$ and $A^t B - B^t A = 0$. These are precisely the symplectic matrices X for which $X^{-1} = X^t$. It follows that the Cartan involution is $\theta : X \mapsto X^{-t}$. The negative eigenspace of the induced map on the Lie agebra $\mathfrak{sp}_{2n}(\mathbb{R})$ is the subspace of matrices such that

$$A = A^t \text{ and } B = -B^t.$$

The Killing form on the Lie algebra is given by

$$(A + iB, C + iD) \mapsto 4\left[n\left\{\operatorname{tr}(AC) - \operatorname{tr}(BD)\right\} - \operatorname{tr}A \cdot \operatorname{tr}B + \operatorname{tr}B \cdot \operatorname{tr}D\right].$$

Some unpleasant algebra verifies that this is invariant under conjugation by elements of K. It is easier to see that it is positive definite on the negative eigenspace of the Cartan involution.

In many cases (whenever the intersection of K with each irreducible factor of G contains a circle in its centre) there is also a G-invariant complex structure with respect to which the metric is Kähler. When this occurs we say D is **Hermitian**. For instance, the above example is Hermitian.

We can embed G into $GL_n(\mathbb{R})$ for some n using a finite-dimensional faithful representation. A discrete subgroup Γ of G is said to be **arithmetic** if there is such an embedding for which Γ is **commensurable** with $G_{\mathbb{Z}} = G \cap GL_n(\mathbb{Z})$, i.e. if the intersection $\Gamma \cap G_{\mathbb{Z}}$ is of finite index in both Γ and $G_{\mathbb{Z}}$. In Example 6.4.1 the group $\Gamma = SP_{2n}(\mathbb{Z})$ is an arithmetic subgroup.

Definition 6.4.2. A **locally symmetric space** is a quotient

$$\Gamma \backslash D = \Gamma \backslash G / K$$

of D by some arithmetic subgroup Γ of G. The metric on D induces a metric on the quotient $\Gamma \backslash D$.

Remarks 6.4.3.   1. In general Γ may act on D with finite stabilisers so that the quotient $\Gamma \backslash D$ is a Riemannian orbifold. This introduces certain technical difficulties but it is not a serious problem; all the results we require go through (with technical modifications) just as they would in the simpler case when Γ acts freely and the quotient is a manifold.

2. The terminology is explained by the following observation. For any point p in a Riemannian manifold M we can find a convex neighbourhood U of $0 \in T_p M$ such that the exponential map

$$\exp : U \to M$$

is a diffeomorphism onto its image. The involution $v \mapsto -v$ of $T_p M$ induces a smooth involutive diffeomorphism of a neighbourhood of p.

When M is a locally symmetric space this local involution is an *isometry* for each point p. In fact, it is possible to show that any Riemannian manifold with this latter property has the form $\Gamma\backslash G/K$ for some Lie group G, compact subgroup K and discrete subgroup Γ (although not all such spaces are locally symmetric in this sense), see e.g. Ji [86].

We say a locally symmetric space $X = \Gamma\backslash D$ is **Hermitian** if D is Hermitian; in this case X clearly inherits a complex structure from D with respect to which its metric is Kähler. In fact a much stronger result is known; every Hermitian locally symmetric space can be given the structure of a quasi-projective variety (Baily and Borel [5]), although it is important to realise that in general the metric does not arise from the restriction of the Fubini–Study metric on projective space. A **locally symmetric variety** is a locally symmetric space considered with this extra structure of a complex variety.

Example 6.4.4. Take $G = SL_2(\mathbb{R})$ and $K = SO_2(\mathbb{R})$. We can identify $D = G/K$ with the upper half-plane

$$\{x + iy \in \mathbb{C} \mid y > 0\}$$

as follows. Recall that $SL_2(\mathbb{R})$ acts transitively on the upper half-plane via fractional linear translations:

$$\begin{pmatrix} a & b \\ c & d \end{pmatrix} : z \longmapsto \frac{az+b}{cz+d}.$$

The stabiliser of the point i under this action is easily seen to be $SO_2(\mathbb{R})$ and hence the orbit of i, which is the whole of the upper half-plane, is naturally identified with $D = G/K$. The natural metric is the **hyperbolic**, or Poincaré, metric

$$\frac{dx^2 + dy^2}{y^2}.$$

Since $SO_2(\mathbb{R}) \cong S^1$ manifestly contains a circle this example is Hermitian; the invariant complex structure is of course the usual one.

Take the arithmetic subgroup Γ to be $SL_2(\mathbb{Z})$. Figure 6.1 shows a fundamental domain for the action of $SL_2(\mathbb{Z})$ on the upper half-plane. The corresponding Hermitian locally symmetric variety X is obtained by glueing the edges of this domain according to the $SL_2(\mathbb{Z})$ action. We can interpret X as the moduli space of elliptic curves.

Example 6.4.5. Here is one way (of several) in which this example generalises to higher dimensions. Take G and K as in Example 6.4.1. Then for $n > 1$ the quotient G/K can be identified with the Siegel upper half-space of genus n consisting of the symmetric complex $n \times n$ matrices with positive definite imaginary part. If we take our arithmetic group to be $\Gamma = SP_{2n}(\mathbb{Z})$ then the associated locally symmetric space is a (coarse) moduli space for principally polarised Abelian varieties of dimension n.

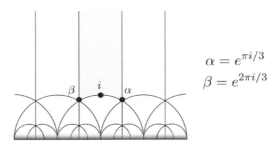

$$\alpha = e^{\pi i/3}$$
$$\beta = e^{2\pi i/3}$$

Figure 6.1: A fundamental domain for the action of $SL_2(\mathbb{Z})$ on the upper half-plane is shown shaded. The other regions are the images of this domain under the action; in this picture the regions become smaller and smaller as we approach the x axis (although they are congruent in the hyperbolic metric).

We can define the L^2-cohomology groups $H^*_{(2)}(X;\mathbb{C})$ of a locally symmetric variety X (with coefficients in $\mathbb{C}$). The metric is complete and so we know that the natural map

$$\mathcal{H}^r(X) \to H^r_{(2)}(X)$$

is injective. If X is Hermitian this map is actually known to be an isomorphism (Zucker [184, p153]). In particular the L^2-cohomology satisfies Poincaré duality and this raises the possibility of a topological interpretation as the intersection cohomology groups of some compactification. (One also obtains a Hodge decomposition on the L^2-cohomology of a locally symmetric variety but that is another story.)

A locally symmetric variety has compactifications called **Satake compactifications** obtained by adding in lower dimensional locally symmetric spaces. In general there is no unique way to do this; there are $2^r - 1$ Satake compactifications where r is the rational rank of G (the rational rank is the dimension of a $\mathbb{Q}$-split torus in the group $G_{\mathbb{Q}} = G \cap GL_n(\mathbb{Q})$). However, when D (and hence X) is Hermitian, there is a distinguished Satake compactification X^* called the **Baily–Borel compactification** which is furthermore a complex projective variety. (Indeed, this is how one proves that Hermitian locally symmetric spaces are varieties.) For example, (as a set) the Baily–Borel compactification of the moduli space of principally polarised Abelian varieties of dimension n is obtained by adding in a boundary component for each moduli space of principally polarised Abelian varieties of dimension $< n$.

If we suppose X is Hermitian then we can consider the natural map

$$H^*_{(2)}(X;\mathbb{C}) \to IH^*(X^*;\mathbb{C}) \tag{6.14}$$

to the intersection cohomology of the Baily–Borel compactfication. More generally, suppose E is an irreducible finite-dimensional complex representation of G. There is a natural action of G on the trivial bundle $D \times E$ on D given by

$$g \cdot (hK, e) = (ghK, \rho(g)e) \qquad (6.15)$$

where $\rho : G \to GL(E)$ is the representation. The quotient $\mathcal{E} = \Gamma \backslash (D \times E)$ is a bundle on $X = \Gamma \backslash D$ again with fibre E. The obvious flat connection on the trivial bundle descends to a flat connection on $\mathcal{E}$ so that we can also think of $\mathcal{E}$ as a local system, or locally constant sheaf.

We can put a Hermitian metric on the bundle $\mathcal{E}$ as follows. There is a Hermitian inner product $\langle \ , \ \rangle$ on the irreducible representation E such that for all $g \in G$ and $e_1, e_2 \in E$ we have

$$\langle e_1, \rho(g)e_2 \rangle = \langle \rho\left(\theta(g)^{-1}\right) e_1, e_2 \rangle$$

where θ is the Cartan involution (Borel and Wallach [23, II.2.2]). Such an inner product is unique up to multiplication by a scalar and is called **admissible**. Note that it is invariant under the action of the compact group K. This inner product gives a Hermitian metric μ on the trivial bundle $D \times E$ with

$$\mu_{gK}(e_1, e_2) = \langle \rho(g^{-1})e_1, \rho(g^{-1})e_2 \rangle.$$

Since $\langle \ , \ \rangle$ is K-invariant this does not depend on the choice of representative of gK. Furthermore this metric is invariant under the action of G on $D \times E$ described in (6.15). Hence it descends to a Hermitian metric on $\mathcal{E}$. Note that, although $\mathcal{E}$ is a flat bundle, this is *not* a flat metric.

We can extend our previous definition to define L^2-cohomology $H^*_{(2)}(X; \mathcal{E})$ with coefficients in the metrised bundle $\mathcal{E}$. Furthermore there is a natural map

$$H^*_{(2)}(X; \mathcal{E}) \to IH^*(X^*; \mathcal{E}) \qquad (6.16)$$

to the intersection cohomology of the Baily–Borel compactification X^* with coefficients in the local system $\mathcal{E}$. When E is the trivial 1-dimensional representation (with the usual Hermitian inner product) we reduce to (6.14).

Theorem 6.4.6 (Zucker's conjecture). *For a Hermitian locally symmetric variety $X = \Gamma \backslash G/K$ and irreducible complex representation E of G the map (6.16) is an isomorphism.*

Zucker made this conjecture in 1980. It was subsequently proved by Looijenga [116] and independently by Saper and Stern [157]. The proofs are quite different; Looijenga's uses an explicit desingularisation of the Baily–Borel compactification X^* whereas Saper and Stern's is Lie-theoretic. More recently Saper has developed a theory of $\mathcal{L}$-modules, a combinatorial analogue of constructible sheaves on the reductive Borel–Serre compactification of a locally symmetric space, which provides yet another proof of Zucker's conjecture.

6.5   Further reading

Cheeger developed the theory of L^2-cohomology in the mid 1970s and first
announced his results in [44], with more details following in [45] and [47].
Cheeger, Goresky and MacPherson [48], Saper [153] and Saper and Zucker
[158] are good overviews of L^2-cohomology and its relation to intersection
cohomology. For a more recent approach to the conjectural isomorphism
between the L^2 and intersection cohomology of a singular projective variety,
using the Riemann–Hilbert correspondence, see Bressler, Saito and Youssin
[34].

The literature on locally symmetric varieties is vast and we make no attempt to survey it here. For an introduction to locally symmetric spaces and
their compactifications see Ji [86] and Borel and Ji [21]. Zucker's paper [184] is
an excellent introduction to the L^2-cohomology of locally symmetric varieties.
His earlier papers [182] and [183] give the context for his conjecture, discuss
its proof in various special cases and also explain the (important) relation
of the conjecture to automorphic forms and the Langlands program. Zucker
[185] is a useful survey of Saper and Stern's and Looijenga's proofs of the conjecture. Saper's article [154] provides a more recent survey of developments
in the area.

In 1983 Borel extended Zucker's conjecture from Hermitian locally symmetric varieties to arithmetic quotients of equal-rank symmetric spaces. In
this situation there is no longer a Baily–Borel compactification. Borel conjectured instead that the intersection cohomology of a Satake compactification
whose boundary components are quotients of equal-rank symmetric spaces
is isomorphic to the L^2-cohomology of the locally symmetric space. This
removes the complex geometry whilst leaving the Lie-theoretic and combinatorial elements of the problem. The methods of Saper and Stern's proof of
Zucker's conjecture can be adapted to this case. Recently Saper has given
a proof based on the theory of $\mathcal{L}$-modules [155]. Saper [154] provides an exposition of the theory of $\mathcal{L}$-modules and also shows how they can be used to
prove Rapoport and Goresky and MacPherson's conjecture that the intersection cohomology of an equal-rank compactification is isomorphic to that of
the reductive Borel–Serre compactification (see also Saper [156]).

Chapter 7

Sheaf-theoretic intersection homology

Let X be a topological pseudomanifold with a fixed topological stratification defined by the filtration

$$X = X_m \supseteq X_{m-1} = X_{m-2} \supseteq \cdots \supseteq X_0.$$

We have defined the intersection homology groups $IH_i(X)$ as the homology groups of a chain complex $IC_\bullet(X)$. In this chapter we shall give a sheaf-theoretic description of $IH_*(X)$ which leads, amongst other things, to a proof that $IH_*(X)$ is a topological invariant of X.

We work with coefficients in a field $\mathbb{F}$.

7.1 Sheaves of singular chains

If U is an open subset of X then $I^p S_\bullet(U)$ is a subcomplex of $I^p S_\bullet(X)$. However the complexes $I^p S_\bullet(U)$ do not define a complex of sheaves on X because if U and V are open subsets such that $V \subset U$ the natural map goes from $I^p S_i(V)$ to $I^p S_i(U)$, not the other way round. To get a complex of sheaves we shall use the complexes $I^p S_i((U))$ of singular intersection chains with closed support instead (see Remark 4.3.2).

Remark 7.1.1. If we are interested in intersection *cohomology* then we can dualise and obtain a complex of presheaves whose sections over U are the singular intersection cochains on U. Using similar techniques to those of this chapter one can show that the hypercohomology groups of the associated complex of sheaves are the intersection cohomology groups of X.

Recall that a formal linear combination

$$\xi = \sum_\sigma \xi_\sigma \sigma$$

of singular i-simplices in X, that is continuous maps from the standard i-simplex $\Delta_i \subset \mathbb{R}^{i+1}$ to X, is a locally finite singular i-chain in $S_i\left(\left(X\right)\right)$ if for each $x \in X$ there is an open neighbourhood U_x of x in X such that the set

$$\{\xi_\sigma | \xi_\sigma \neq 0, \sigma^{-1}(U_x) \neq \emptyset\}$$

is finite. It is a singular i-chain in $S_i(X)$ if $\{\xi_\sigma | \xi_\sigma \neq 0\}$ is finite.

Suppose $V \subset U$ are open subsets of X. We now define a restriction map

$$S_i\left(\left(U\right)\right) \to S_i\left(\left(V\right)\right) : \xi \mapsto \xi|_V.$$

First note that it is sufficient to define the restriction for a single singular i-simplex σ in U and then extend linearly. Given σ we define a set J of singular i-simplices in V as follows:

- If $\operatorname{im} \sigma \subset V$ then put $J = \{\sigma\}$;

- Otherwise perform a barycentric subdivision of σ (see e.g. Hatcher [80, §2.1]). If τ is an i-simplex in the subdivision with $\operatorname{im} \tau \subset V$ then add τ to J. Further subdivide those i-simplices in the subdivision with $\operatorname{im} \tau \not\subset V$ and repeat.

In this way we obtain a well-defined set J of singular i-simplices in V. We define the restriction by

$$\sigma|_V = \sum_{\sigma \in J} \sigma.$$

This is a locally-finite i-chain in V with the property that its support is $|\sigma| \cap V$. We can now extend linearly to define the restriction $S_i\left(\left(U\right)\right) \to S_i\left(\left(V\right)\right)$.

Thus we obtain a presheaf $U \mapsto S_i\left(\left(U\right)\right)$ on X which we can verify is in fact a sheaf. Note that the restriction commutes with taking boundaries, i.e. $(\partial \xi)|_V = \partial(\xi|_V)$ so that the boundary operator on chains induces a sheaf map. By convention, because people like to work with cochain complexes of sheaves, not chain complexes, the sheaf defined by $U \mapsto S_i\left(\left(U\right)\right)$ is denoted by $\mathcal{S}_X^{-i}$ so that the boundary induces a sheaf map $\partial : \mathcal{S}_X^{-i} \to \mathcal{S}_X^{-i+1}$.

The same construction works for intersection chains. A locally finite singular i-chain ξ is in the subspace $I^p S_i\left(\left(U\right)\right)$ if it is p-allowable (see Definition 4.3.1). If ξ is p-allowable then so is its restriction $\xi|_V$ to an open subset V of U. It follows that we can define a complex of sheaves $\mathcal{I}^p \mathcal{S}_X^\bullet$ with

$$\mathcal{I}^p \mathcal{S}_X^{-i}(U) = I^p S_i\left(\left(U\right)\right).$$

Warning 3. Unfortunately there is inconsistency in the literature in the indexing of the sheaf complex $\mathcal{I}^p \mathcal{S}_X^\bullet$. The only consistency is in working with sheaves of cochain complexes and not with sheaves of chain complexes. Sometimes the index $-i$ is replaced by $m - i$ or $\frac{m}{2} - i$ if X is a complex variety (because $\frac{m}{2}$ is then the complex dimension of X). See Goresky and MacPherson [70, §2.3].

Remark 7.1.2. The sections $\mathcal{F}(U)$ of a sheaf $\mathcal{F}$ over an open subset U are often denoted $\Gamma(U; \mathcal{F})$ and the subspace of sections with compact support by $\Gamma_c(U; \mathcal{F})$. In particular $\Gamma(U; \mathcal{I}^p\mathcal{S}_X^{-i}) = I^pC^i\left((U)\right)$ and the sections with *compact* support are the finite chains, i.e.

$$\Gamma_c(U; \mathcal{I}^p\mathcal{S}_X^{-i}) = IC_i(U).$$

Under this identification the sheaf map $\partial \colon \mathcal{I}^p\mathcal{S}_X^{-i} \to \mathcal{I}^p\mathcal{S}_X^{-i+1}$ induces the original boundary map $\partial \colon I^pS_i(U) \to I^pS_{i-1}(U)$.

Remark 7.1.3. If $\mathcal{L}$ is a local coefficient system over X then we can define complexes of sheaves $\mathcal{S}_{(X,\mathcal{L})}^\bullet$ and $\mathcal{I}^p\mathcal{S}_{(X,\mathcal{L})}^\bullet$ over X in the obvious way (cf. Section 4.9). Indeed to define $\mathcal{I}^p\mathcal{S}_{(X,\mathcal{L})}^\bullet$ we only need a local coefficient system $\mathcal{L}$ over the non-singular open subset $X - X_{m-2}$ of X, not over X itself.

Our aim is to compute the hypercohomology of the complexes $\mathcal{S}_X^\bullet$ and $\mathcal{I}^p\mathcal{S}_X^\bullet$. The first step is to show that the higher cohomology groups of the sheaves $\mathcal{S}_X^i$ and $\mathcal{I}^p\mathcal{S}_X^i$ vanish.

Definition 7.1.4. A sheaf $\mathcal{F}$ on X is **soft** if for every closed subset $A \subset X$ the restriction map

$$\mathcal{F}(X) \to \mathcal{F}(A)$$

is surjective. Here, by definition, $\mathcal{F}(A) = \mathrm{colim}_{U \supset A}\mathcal{F}(U)$ is the colimit over open subsets U containing A (which can be thought of as the 'stalk' of $\mathcal{F}$ at the closed subset A). A sheaf $\mathcal{F}$ on X is **c-soft** if for every compact subset $K \subset X$ the restriction map $\mathcal{F}(X) \to \mathcal{F}(K)$ is surjective.

In fact, since topological pseudomanifolds are locally compact and countable at infinity, c-soft sheaves on X are soft (the converse is clear) — see Kashiwara and Schapira [97, Ex. II.6]. Soft sheaves are cohomologically trivial, in other words if $\mathcal{F}$ is a soft sheaf then $H^p(X; \mathcal{F}) = 0$ for $p > 0$, see Iversen [84, p157].

Lemma 7.1.5. *The sheaves $\mathcal{S}_X^i$ and $\mathcal{I}^p\mathcal{S}_X^i$ are c-soft (and hence by the above are soft). In particular, for each i and for $p > 0$ we have*

$$H^p(X; \mathcal{S}_X^i) = 0 = H^p(X; \mathcal{I}^p\mathcal{S}_X^i).$$

Proof. Let $K \subset X$ be compact and suppose $\zeta \in \mathcal{S}_X^i(K)$. We can represent ζ by a locally finite singular chain $\xi \in \mathcal{S}_X^i(U)$ for some open neighbourhood U of K. Every point $x \in K$ has a neighbourhood V_x which meets only finitely many singular simplices of ξ. The V_x form an open cover of K; take a finite subcover. There are finitely many simplices $\{\sigma_j | j \in F\}$ of ξ meeting the open sets in this subcover, so we have a (finite) chain

$$\tilde{\xi} = \sum_{j \in F} \xi_{\sigma_j}\sigma_j \in \mathcal{S}_X^i(U)$$

and composing with the inclusion map $U \hookrightarrow X$ we can think of $\tilde{\xi} \in \mathcal{S}_X^i(X)$. Since every singular simplex of ξ which meets K is in $\{\sigma_j | j \in F\}$ we have $\tilde{\xi}|_K = \xi|_K = \zeta$.

The argument for $\mathcal{I}^p\mathcal{S}_X^i$ is the same. $\square$

Suppose $\mathcal{F}$ is a soft sheaf on X. Since its higher cohomology groups vanish the complex $p_{X*}\mathcal{I}^\bullet(\mathcal{F})$ of global sections of the Godement resolution must be exact except at the degree zero term. It follows that the quasi-isomorphism $\mathcal{F} \to \mathcal{I}^\bullet(\mathcal{F})$ of $\mathcal{F}$ with its Godement resolution induces a quasi-isomorphism

$$p_{X*}\mathcal{F} \to p_{X*}\mathcal{I}^\bullet(\mathcal{F})$$

on global sections. The double complex theorem (see the proof of Proposition 3.5.8) allows us to generalise this to complexes: if $\mathcal{F}^\bullet$ is a complex of soft sheaves then there is a quasi-isomorphism $p_{X*}\mathcal{F}^\bullet \to p_{X*}\mathcal{I}^\bullet(\mathcal{F})$. This immediately implies

$$H^i(p_{X*}\mathcal{F}^\bullet) \cong H^i(p_{X*}\mathcal{I}^\bullet(\mathcal{F})) \cong H^i(X; \mathcal{F}^\bullet),$$

i.e. the hypercohomology of $\mathcal{F}^\bullet$ is isomorphic to the cohomology of the complex of global sections of $\mathcal{F}^\bullet$. The important examples for us are

$$H_i^{cl}(X) = H^{-i}(p_{X*}\mathcal{S}_X^\bullet) \cong H^{-i}(X; \mathcal{S}_X^\bullet)$$

and

$$\mathcal{I}^p H_i^{cl}(X) = H^{-i}(p_{X*}\mathcal{I}^p\mathcal{S}_X^\bullet) \cong H^{-i}(X; \mathcal{I}^p\mathcal{S}_X^\bullet).$$

The same arguments apply to sections with compact support and so we also have

$$H_i(X) = H^{-i}(p_{X!}\mathcal{S}_X^\bullet) \cong H_c^{-i}(X; \mathcal{S}_X^\bullet)$$

and

$$\mathcal{I}^p H_i(X) = H^{-i}(p_{X!}\mathcal{I}^p\mathcal{S}_X^\bullet) \cong H_c^{-i}(X; \mathcal{I}^p\mathcal{S}_X^\bullet).$$

Remark 7.1.6. In the first edition of this book sheaves of locally finite *simplicial* chains $\mathcal{C}_X^i$ and $\mathcal{I}\mathcal{C}_X^i$ were used. Section 5.2 proved that $\mathcal{C}_X^i$ was a **fine** sheaf (see Remark 5.2.4 in the first edition) and it was asserted, following Goresky and MacPherson [70, §2.1], that $\mathcal{I}\mathcal{C}_X^i$ was also fine. However, Habegger pointed out in [24, II §5] that this is not immediately clear since, unlike ordinary simplicial chains, intersection chains cannot be broken into pieces arbitrarily (because this may create non-allowable boundary components). Fortunately the sheaves $\mathcal{I}\mathcal{C}_X^i$ are soft (Borel *et al.* [24, II §5]) and this is quite sufficient for our purposes.

7.2   Constructibility and an axiomatic characterisation

Let X be a topological m-pseudomanifold. We fix a stratification $X = X_m \supset X_{m-1} = X_{m-2} \supset \cdots \supset X_0$ of X and label it by $\mathbb{S}$.

In the previous section we showed that the closed and compact support intersection homology groups are respectively the hypercohohomology and compactly supported hypercohomology of a certain bounded complex of sheaves $\mathcal{I}^p \mathcal{S}_X^\bullet$. Put another way, this complex determines the groups. Since we are really only interested in the groups up to isomorphism, we are only interested in the complex of sheaves up to a sequence of quasi-isomorphisms or, equivalently, up to an isomorphism in the bounded derived category $D^b(X)$ of sheaves on X.

In this section we give a set of conditions or axioms on a complex of sheaves which are satisfied by $\mathcal{I}^p \mathcal{S}_X^\bullet$. In fact, although we will not prove it, these conditions determine $\mathcal{I}^p \mathcal{S}_X^\bullet$ up to isomorphism in $D^b(X)$. Before we give the axioms we need to introduce some terminology.

Constructibility

Definition 7.2.1. A sheaf $\mathcal{F}$ on X is **constructible** with respect to the stratification $\mathbb{S}$, or $\mathbb{S}$-constructible, if the restriction

$$\mathcal{F}|_{X_{m-k} - X_{m-k-1}}$$

is a locally constant sheaf on $X_{m-k} - X_{m-k-1}$ with finite-dimensional stalk for each $k \geq 0$.

A bounded complex $\mathcal{F}^\bullet$ of sheaves is **cohomologically $\mathbb{S}$-constructible** if the cohomology sheaves $\mathcal{H}^i(\mathcal{F}^\bullet)$ are constructible for all i.

Remark 7.2.2. The restriction $F|_Y$ of a sheaf F on X to a subspace Y is the sheaf on Y defined as follows. If U is an open subset of Y then $F|_Y(U)$ is the colimit with respect to the restriction of the vector spaces $F(V)$ for V open in X such that $U \subseteq V$.

It is convenient to consider the full subcategory of complexes in $D^b(X)$ which are cohomologically $\mathbb{S}$-constructible (that is, the category whose objects are bounded complexes of sheaves on X which are cohomologically $\mathbb{S}$-constructible and with the maps between any two such being the same as in $D^b(X)$). We denote this by $D_\mathbb{S}^b(X)$ and refer to it as the $\mathbb{S}$-**constructible bounded derived category of** X. It is a triangulated category (Gelfand and Manin [63, Ch. 7 §1.6] or Dimca [56, Ch. 4]).

Remark 7.2.3. More generally, if we specify a collection A of stratifications of X such that any two stratifications in A have a common refinement in A then the full subcategory $D_A^b(X)$ of complexes which are cohomologically constructible with respect to some stratification in A is triangulated.

Taking $A = \{\mathbb{S}\}$ we recover the previous definition but there are other important examples. If X is a complex algebraic variety then we can consider the algebraically constructible derived category $D^b_{\text{alg}-c}(X)$ where we take A to be the collection of Whitney stratifications by algebraic subvarieties. In a similar vein we can consider the analytically constructible derived category $D^b_{\text{an}-c}(X)$ of a complex analytic variety or the PL constructible derived category $D^b_{\text{PL}-c}(X)$ of a piecewise-linear pseudomanifold. See Dimca [56, Ch. 4] for more details.

Stalks and costalks

Let $\jmath_x : \{x\} \hookrightarrow X$ be the inclusion of a point x. Recall that the **stalk** at x of a sheaf $\mathcal{E}$ is the vector space

$$\jmath_x^* \mathcal{E} = \operatorname{colim}_{U \ni x} \Gamma(U; \mathcal{E}).$$

The functor $\jmath_x^*$ is exact and so we can extend this to a functor $D^b(X) \to D^b(x)$ on derived categories simply by applying $\jmath_x^*$ term-by-term to a complex. For a bounded complex $\mathcal{F}^\bullet$ of sheaves on X we will refer to the resulting complex $\jmath_x^* \mathcal{F}^\bullet$ of vector spaces as the **stalk** of $\mathcal{F}^\bullet$ at x.

There is a complementary construction called the **costalk**. If $V \subset U$ are nested neighbourhoods of x then a section which is compactly supported in V can be extended by zero (remember sections of sheaves which agree on overlaps can always be patched together) to obtain a compactly supported section over U. In other words 'extension by zero' defines a map

$$\Gamma_c(V; \mathcal{E}) \to \Gamma_c(U; \mathcal{E}).$$

The **costalk** at x of a sheaf $\mathcal{E}$ on X is the limit

$$\Gamma_x \mathcal{E} = \lim_{U \ni x} \Gamma_c(U; \mathcal{F}^\bullet).$$

The functor Γ_x is left exact (Iversen [84, II Prop. 6.8]) and so there is a right derived functor

$$R\Gamma_x : D^b(X) \to D^b(x).$$

Remark 7.2.4. We will actually use the notation $R\Gamma_x = \jmath_x^!$ because this is a special case of a more general functor $f^! : D^b(Y) \to D^b(X)$ which is associated to a continuous map $f : X \to Y$ between finite-dimensional locally compact spaces — see Iversen [84, VIII Thm. 3.1].

Although we will neither use nor emphasise this, the functor $f^!$ can be described as $D_X f^* D_Y$ where D_X and D_Y are the Verdier duality functors on $D^b(X)$ and $D^b(Y)$ respectively (see §7.4). This is a sheaf-theoretic generalisation of the following construction of a map between the homology groups of manifolds: if $f : X \to Y$ is a map of manifolds then we define $f^!$ by

$$
\begin{array}{ccc}
H_i(Y) & \xrightarrow{\hspace{2cm}} & H^{\dim X - i}(Y) \\
{\scriptstyle f^!} \downarrow & & \downarrow {\scriptstyle f^*} \\
H_{\dim Y - \dim X + i}(X) & \xleftarrow{\hspace{2cm}} & H^{\dim X - i}(X)
\end{array}
$$

where the horizontal maps are the isomorphisms given by Poincaré duality.

For a bounded complex $\mathcal{F}^\bullet$ we will refer to the complex $j_x^! \mathcal{F}^\bullet$ of vector spaces as the **costalk** at x.

The axioms

We say a complex of sheaves $\mathcal{E}^\bullet \in D^b(X)$ satisfies $\mathrm{AX_p[\mathbb{S}]}$ if it is cohomologically $\mathbb{S}$-constructible, in other words it is in the subcategory $D^b_{\mathbb{S}}(X)$, and

(a) for any $x \in X$ the stalk cohomology $H^i(j_x^* \mathcal{E}^\bullet) = 0$ for $i < -m$;

(b) for any $x \in X - X_{m-2}$ the stalk cohomology satisfies

$$
H^i(j_x^* \mathcal{E}^\bullet) \cong \begin{cases} \mathbb{F} & i = -m \\ 0 & \text{otherwise,} \end{cases}
$$

and, furthermore, the $(-m)$th stalk cohomology forms the constant local system on $X - X_{m-2}$;

and, for each x in a codimension $k > 0$ stratum, i.e. $x \in X_{m-k} - X_{m-k-1}$

(c) the ith stalk cohomology $H^i(j_x^* \mathcal{E}^\bullet) = 0$ for $i > p(k) - m$;

(d) the ith costalk cohomology $H^i(j_x^! \mathcal{E}^\bullet) = 0$ for $i < -q(k)$.

Here q is the complementary perversity to p, i.e. $p(k) + q(k) = k - 2$.

These axioms are equivalent to the reformulation AX1[d''] of AX1 in Goresky and MacPherson [70, §3.3]. We sketch a proof that they are satisfied by $\mathcal{I}^p \mathcal{S}_X^\bullet$ below. The full details of the simplicial case, which is entirely analogous, are in Goresky and MacPherson [70].

Recall that a neighbourhood of a point in X is modelled on a cone and we have vanishing results (see §4.7) for the intersection homology of a cone. In a little more detail, suppose S is a stratum of codimension k, i.e. a connected component of $X_{m-k} - X_{m-k-1}$. Recall that there is a topological pseudomanifold L_S, the link of S, of dimension $k - 1$ such that for any $x \in S$ there is a neighbourhood N_x of x in X and a homeomorphism

$$
N_x \cong \mathbb{R}^{m-k} \times C(L_S).
$$

(We allow the case $k = 0$ for which $L_S = \emptyset$ and $C(L_S)$ is a single point.) We have (using the fact that taking cohomology commutes with taking stalks)

$$
H^i(j_x^* \mathcal{I}^p \mathcal{S}_X^\bullet) \cong \mathcal{I}^p H^{cl}_{-i}(N_x).
$$

Axiom (a) is satisfied because $\dim N_x = m$. Axiom (b) is satisfied because any $x \in X - X_{m-2}$ has a neighbourhood $N_x \cong \mathbb{R}^m$ and

$$I^p H^{cl}_j(\mathbb{R}^m) \cong H^{cl}_j(\mathbb{R}^m) \cong \left\{ \begin{array}{ll} \mathbb{F} & j = m \\ 0 & \text{otherwise.} \end{array} \right. \tag{7.1}$$

Note that for $x \in X_{m-k} - X_{m-k-1}$ we have

$$I^p H^{cl}_{-i}(N_x) \cong I^p H^{cl}_{-i}(\mathbb{R}^{m-k} \times C(L_S)) \cong I^p H^{cl}_{k-m-i}(C(L_S))$$

by a version of the Künneth theorem for closed support intersection homology. So axiom (c) holds because the closed support intersection homology of a cone vanishes in 'low' dimensions (see §4.7). It also follows that $I^p S^{\bullet}_X$ is cohomologically $\mathbb{S}$-constructible.

The cohomology of the costalk at a point $x \in X_{m-k} - X_{m-k-1}$ turns out to be

$$H^i(j_x^! I^p S^{\bullet}_X) \cong I^p H_{-i}(C(L_S)).$$

So axiom (d) holds because the intersection homology of a cone vanishes in 'high' dimensions (see §4.7) and we are done.

Not only does $I^p S^{\bullet}_X$ satisfy these axioms, it is determined by them up to unique isomorphism in $D^b_{\mathbb{S}}(X)$. The proof in Goresky and MacPherson [70, §3.5] proceeds inductively beginning with $X - X_{m-2}$ and adding strata in order of increasing codimension. To recapitulate

Theorem 7.2.5 (cf. Goresky and MacPherson [70, §3.5]). *Up to canonical isomorphism in the $\mathbb{S}$-constructible bounded derived category $D^b_{\mathbb{S}}(X)$ there is a unique complex of sheaves $\mathcal{E}^{\bullet}$ which satisfies the axioms $AX_p[\mathbb{S}]$. Since $I^p S^{\bullet}_X$ satisfies the axioms it follows that there are canonical isomorphisms*

$$H^i(X; \mathcal{E}^{\bullet}) \cong IH^{cl}_{-i}(X) \quad \text{and} \quad H^i_c(X; \mathcal{E}^{\bullet}) \cong IH_{-i}(X).$$

Remark 7.2.6. Suppose $\mathcal{L}$ is a local system on $X - X_{m-2}$. Then, with the obvious modification to (b), these axioms characterise $I^p S_{X,\mathcal{L}}$ up to isomorphism in $D^b_{\mathbb{S}}(X)$.

7.3   The topological invariance of intersection homology

Following Goresky and MacPherson [70, §4.1] we can recast the axioms $AX_p[\mathbb{S}]$ in a stratification-independent form. We say a complex $\mathcal{E}^{\bullet}$ of sheaves in the bounded derived category $D^b(X)$ of a topological pseudomanifold X satisfies AX_p if

(a) for any $x \in X$ the stalk cohomology $H^i(j_x^* \mathcal{E}^{\bullet}) = 0$ for $i < -m$;

(b) there is a closed subset $\Sigma \subset X$ of codimension ≥ 2 such that for any $x \in X - \Sigma$ the stalk cohomology satisfies

$$H^i(\jmath_x^* \mathcal{E}^\bullet) \cong \begin{cases} \mathbb{F} & i = -m \\ 0 & \text{otherwise,} \end{cases}$$

and, furthermore, the $(-m)$th stalk cohomology forms the constant local system on $X - \Sigma$;

and for $i > 0$ we have the following restrictions on the codimension of the sets on which the $(-i)$th cohomology of the stalk and costalk are non-vanishing:

(c) $\operatorname{codim} \{H^{-i}(\jmath_x^* \mathcal{E}^\bullet) \neq 0\} \geq p^{-1}(m - i)$

(d) $\operatorname{codim} \{H^{-i}(\jmath_x^! \mathcal{E}^\bullet) \neq 0\} \geq q^{-1}(i)$

where, as before, q is the complementary perversity to p.

Remarks 7.3.1.   1. Here by p^{-1} we mean the subinverse, i.e.

$$p^{-1}(n) = \min\{a | p(a) \geq n\},$$

and similarly for q^{-1}.

2. To make sense of these axioms we need a notion of dimension for subsets of a topological space; Goresky and MacPherson use that of Hurewicz and Wallman [83] in [70].

3. As with $\mathrm{AX_p}[\mathbb{S}]$, we can modify (b) so that the modified axioms characterise $\mathcal{I}^p \mathcal{S}_{X,\mathcal{L}}$ where $\mathcal{L}$ is a local system on $X - \Sigma$.

Exercise 7.3.2. A complex satisfies $\mathrm{AX_p}[\mathbb{S}]$ if and only if it satisfies $\mathrm{AX_p}$ and is cohomologically $\mathbb{S}$-constructible.

We want to show that intersection homology is a topological invariant, i.e. that a homeomorphism $f \colon X \to Y$ induces an isomorphism

$$f_* \colon IH_*(X) \to IH_*(Y).$$

In order to show this it now suffices to find a complex $\mathcal{P}^\bullet$ which satisfies the stratification-independent axioms $\mathrm{AX_p}$ *and* which is cohomologically $\mathbb{S}$-constructible with respect to *any* stratification $\mathbb{S}$. For it then follows from Exercise 7.3.2 that $\mathcal{P}^\bullet$ satisfies $\mathrm{AX_p}[\mathbb{S}]$ for any topological stratification, and so by Theorem 7.2.5 it is canonically isomorphic in $D^b(X)$ to the complex $\mathcal{I}^p \mathcal{S}_X^\bullet$ defined with respect to that stratification. These isomorphisms induce canonical isomorphisms between intersection homology groups (with both closed and compact supports) defined with respect to different stratifications.

To carry out this programme we need two ideas, Deligne's construction and the canonical filtration.

Deligne's construction

Suppose we have a fixed filtration

$$X = X_m \supseteq X_{m-2} \supseteq \cdots \supseteq X_0$$

of X by closed subsets, which we denote by $\mathbb{S}$. Note that we do not require that this is a stratification, or even that $X_i - X_{i-1}$ is a manifold. Let

$$\imath_k \colon X - X_{m-k} \to X - X_{m-k-1}$$

be the inclusion. Recall from Section 3.4 that if $\mathcal{F}$ is a sheaf on $X - X_{m-k}$ then $\imath_{k*}\mathcal{F}$ is the sheaf on $X - X_{m-k-1}$ satisfying

$$(\imath_{k*}\mathcal{F})(V) = \mathcal{F}(V \cap (X \quad X_{m-k})) = \mathcal{F}\left(\imath_k^{-1}(V)\right) \qquad (7.2)$$

for any open subset V of $X - X_{n-k-1}$.

Definition 7.3.3. If $\mathcal{E}^\bullet$ is a complex of sheaves on X and $r \in \mathbb{Z}$ we define the **truncated complex** $\tau_{\leq r}\mathcal{E}^\bullet$ to be the complex which in degree i is

$$\begin{cases} \mathcal{E}^i & i < p, \\ \ker\left(d \colon \mathcal{E}^r \to \mathcal{E}^{r+1}\right) & i = p, \\ 0 & i > p. \end{cases}$$

Theorem 7.3.4 (Deligne's construction, Goresky and MacPherson [70, §3]). *If $\mathbb{S}$ is a topological stratification then the complex of sheaves*

$$\mathcal{P}^\bullet(\mathbb{S}) = \tau_{\leq p(m)-m} R\imath_{m*} \cdots \tau_{\leq p(2)-m} R\imath_{2*} \mathbb{F}_{X-X_{m-2}}[m]$$

satisfies $\mathrm{AX}_\mathrm{p}[\mathbb{S}]$, *and hence is isomorphic to* $\mathcal{I}^p\mathcal{S}_X^\bullet$. *(As in Section 3.7 $R\imath_{k*}$ is the right derived functor of $\imath_{k*}$.)*

   This construction is very important because it can be used to define intersection homology in any situation where there is a good sheaf theory and good stratifications, for example algebraic geometry in characteristic $p > 0$ (see Chapter 10). Moreover, this construction does not require the filtration of X to be a topological stratification.

The canonical filtration

There is a 'canonical filtration'

$$X = X_m^{\mathrm{can}} \supseteq X_{m-2}^{\mathrm{can}} \supseteq \cdots \supseteq X_0^{\mathrm{can}}$$

of X by closed subsets X_j^{can} which is coarser than any topological stratification in the sense that $X_j^{\mathrm{can}} \subseteq X_j$ for any topological stratification

$$X = X_m \supseteq X_{m-2} \supseteq f \supseteq X_0.$$

The $X_j^{\mathrm{can}} - X_{j-1}^{\mathrm{can}}$ will *not* necessarily be manifolds but they *will* be uniquely determined by X without the need of choice.

Definition 7.3.5. Let U be the largest open subset of X such that the co-homology sheaves of the complex $\mathcal{S}_X^\bullet$ restricted to U are all locally constant. (Equivalently U is the union of all open subsets of X upon which the restriction of the cohomology sheaves of $\mathcal{S}_X^\bullet$ are locally constant). Let X_{m-2}^{can} be the closed subset $X - U$ of X.

If $x \in X$ is not in the singular set X_{m-2} of some topological stratification $X \supseteq X_{m-2} \supseteq \cdots \supseteq X_0$ of X then the ith cohomology stalk of $\mathcal{S}_X^\bullet$ at x is given by

$$H^i(\jmath_x^* \mathcal{S}_X^\bullet) = H_{-i}^{cl}(\mathbb{R}^m) = \begin{cases} \mathbb{F} & i = -m \\ 0 & \text{otherwise.} \end{cases}$$

It follows that

$$X_{m-2}^{\mathrm{can}} \subseteq X_{m-2}$$

and so has codimension ≥ 2. Define X_{m-k}^{can} inductively for $2 < k < m$ as follows. If $1 \leq j \leq k$ let

$$\imath_j^{\mathrm{can}} : X - X_{m-j}^{\mathrm{can}} \to X - X_{m-j-1}^{\mathrm{can}}$$

and

$$\jmath_k : X - X_{m-k}^{\mathrm{can}} \to X$$

be the inclusions. Then let X_{m-k-1}^{can} be the complement in X_{m-k}^{can} of the largest open subset of X_{m-k}^{can} on which the cohomology sheaves of both

$$\mathcal{S}_{X_{m-k}^{\mathrm{can}}}^\bullet \quad \text{and} \quad R(\jmath_k)_* \tau_{\leq p(k-1)-m} R\left(\imath_{k-1}^{\mathrm{can}}\right)_* \cdots R\left(\imath_2^{\mathrm{can}}\right)_* \tau_{\leq p(1)-m} R\left(\imath_1^{\mathrm{can}}\right)_* \mathcal{S}_U^\bullet$$

are locally constant. By induction each X_{m-k}^{can} is a closed subset of X and the filtration

$$X = X_m^{\mathrm{can}} \supseteq X_{m-2}^{\mathrm{can}} \supseteq \cdots \supseteq X_0^{\mathrm{can}}$$

is coarser than any topological stratification

$$X = X_m \supseteq X_{m-2} \supseteq \cdots \supseteq X_0$$

(in the sense that $X_j^{\mathrm{can}} \subseteq X_j$ for all $0 \leq j \leq m$).

Deligne's construction can be applied to the canonical filtration to give a complex $\mathcal{P}^\bullet$ which satisfies $\mathrm{AX_p}$ and furthermore, because the canonical filtration is coarser, $\mathcal{P}^\bullet$ is cohomologically $\mathbb{S}$-constructible for *any* topological stratification $\mathbb{S}$. As noted on page 111 it follows that intersection homology is a topological invariant.

7.4 Duality in the derived category

One of the most important properties of vector spaces is that there is a notion of dual vector space. More formally there is a contravariant functor $V \mapsto \mathrm{Hom}(V, \mathbb{F})$ from the (Abelian) category of vector spaces over $\mathbb{F}$ to itself which takes a vector space to its dual. It is natural to ask whether there is a similar notion of dual for sheaves of vector spaces on a space X. The naive attempt to define the dual of a sheaf $\mathcal{E}$ by

$$U \mapsto \mathcal{E}(U)^{\vee} = \operatorname{Hom}(\mathcal{E}(U), \mathbb{F})$$

for open $U \subseteq X$ fails. This is because there is no natural way to make this into a sheaf, i.e. to define restriction maps: the duals of the restriction maps of $\mathcal{E}$ go in the wrong direction.

A more sensible (categorical) approach is to recall that for any sheaf $\mathcal{F}$ on X there is a functor $\mathcal{H}om_{\operatorname{Sh}(X)}(-, \mathcal{F})$ which takes a sheaf $\mathcal{E}$ on X to the sheaf with sections over an open subset U of X given by the space of sheaf maps from the restriction of $\mathcal{E}$ to the restriction of $\mathcal{F}$, i.e.

$$\mathcal{H}om_{\operatorname{Sh}(X)}(\mathcal{E}, \mathcal{F})(U) = \operatorname{Hom}_{\operatorname{Sh}(U)}(\mathcal{E}|_U, \mathcal{F}|_U).$$

By analogy with the situation for vector spaces we might hope that there is a sheaf $\mathcal{D}_X$ on X, a *dualising sheaf*, for which the assignment

$$\mathcal{E} \mapsto \mathcal{H}om_{\operatorname{Sh}(X)}(\mathcal{E}, \mathcal{D}_X)$$

is an appropriate dual. This is not the case and the reason is that for non-0-dimensional spaces the dualising object is not a sheaf (which we should think of as a complex concentrated in degree zero) but a *complex* of sheaves. Thus we are forced to think in terms of complexes of sheaves, more precisely in terms of the derived category, rather than of sheaves. (Indeed this is a strong motivation for introducing the derived category.) Specifically, if X is a topological pseudomanifold then the contravariant functor

$$D_X(-) = R\mathcal{H}om_{\operatorname{Sh}(X)}(-, \mathcal{S}_X^{\bullet}) : D^b(X) \to D^b(X),$$

is called the **Verdier dual**. Here $\mathcal{S}_X^{\bullet}$ is the sheaf complex of singular chains with closed support (see p104) and $R\mathcal{H}om_{\operatorname{Sh}(X)}(-, \mathcal{S}_X^{\bullet})$ is the right derived functor of the left exact functor $\mathcal{H}om_{\operatorname{Sh}(X)}(-, \mathcal{S}_X^{\bullet})$.

Theorem 7.4.1 (See Gelfand and Manin [63, Ch. 4 §5]). *Suppose X is a topological pseudomanifold. Then the Verdier dual $D_X : D^b(X) \to D^b(X)$ is a contravariant functor such that*

1. *D_X is triangulated (in the contravariant sense), i.e. D_X takes distinguished triangles to distinguished triangles and*

$$D_X\left(\mathcal{E}^{\bullet}[1]\right) = \left(D_X\mathcal{E}^{\bullet}\right)[-1];$$

2. *when X is a point the Verdier dual D_{pt} is (isomorphic to) the standard dual of a complex of vector spaces, i.e. $D_{\mathrm{pt}}V^{\bullet}$ is the complex with $\left(V^{-i}\right)^{\vee}$ in degree i and with differentials given by the duals of the differentials of $V^{\bullet}$;*

3. *for any $\mathcal{E}^\bullet \in D^b(X)$ there is a natural map $\mathcal{E}^\bullet \to D_X^2 \mathcal{E}^\bullet$;*

4. *the Verdier dual commutes with restriction to open subsets, i.e.*

$$D_U(\mathcal{E}^\bullet|_U) = (D_X\mathcal{E}^\bullet)|_U$$

for any complex $\mathcal{E}^\bullet$ of sheaves on X;

5. *[Verdier duality] for any continuous map $f : X \to Y$ and $\mathcal{E}^\bullet$ in $D^b(X)$, there is a natural isomorphism,*

$$D_Y Rf_*\mathcal{E}^\bullet \cong Rf_! D_X\mathcal{E}^\bullet;$$

6. *the Verdier dual preserves cohomological constructibility, i.e. it descends to a contravariant functor $D_X : D_\mathbb{S}^b(X) \to D_\mathbb{S}^b(X)$ for any topological stratification $\mathbb{S}$ of X. Furthermore, the natural map $\mathcal{E}^\bullet \to D_X^2\mathcal{E}^\bullet$ is an isomorphism for a cohomologically constructible complex $\mathcal{E}^\bullet$.*

Remark 7.4.2. If $\mathcal{D}_X^\bullet$ is a complex of sheaves on X for which the functor $R\mathcal{H}om_{\mathrm{Sh}(X)}(-, \mathcal{D}_X^\bullet)$ satisfies the above properties then, since we noted in (3.5) that $\mathcal{H}om_{\mathrm{Sh}(X)}(\mathbb{F}_X, \mathcal{E}) \cong \mathcal{E}$ for any sheaf $\mathcal{E}$, we find that

$$\mathcal{D}_X^\bullet \cong R\mathcal{H}om_{\mathrm{Sh}(X)}(\mathbb{F}_X, \mathcal{D}_X^\bullet)$$

is the 'dual' of the constant sheaf $\mathbb{F}_X$. Hence, using Verdier duality,

$$
\begin{aligned}
H^{-i}(X; \mathcal{D}_X^\bullet) &= H^{-i}(Rp_{X*}\mathcal{D}_X^\bullet) \\
&\cong H^{-i}(D_{\mathrm{pt}}Rp_{X!}\mathbb{F}_X) \\
&\cong H^i(Rp_{X!}\mathbb{F}_X)^\vee \\
&\cong H_c^i(X)^\vee
\end{aligned}
$$

is dual to the compactly supported cohomology of X. A local version of this argument shows that the above properties force the 'dualising complex' to be the sheaf complex $\mathcal{S}_X^\bullet$ of singular chains with closed support.

Verdier duality relates the hypercohomology of the dual $D_X\mathcal{E}^\bullet$ to the compactly supported hypercohomology of $\mathcal{E}^\bullet$ because

$$
\begin{aligned}
H^i(U; D_X\mathcal{E}^\bullet) &= H^i(Rp_{U*}D_U\mathcal{E}^\bullet) \\
&\cong H^i(D_{\mathrm{pt}}Rp_{U!}\mathcal{E}^\bullet) \\
&\cong H^{-i}(Rp_{U!}U)^\vee \\
&\cong H_c^{-i}(U; \mathcal{E}^\bullet)^\vee.
\end{aligned}
$$

An important consequence is that $\mathcal{E}^\bullet$ satisfies $\mathrm{AX}_\mathrm{p}(c)$ and $\mathrm{AX}_\mathrm{p}(d)$ if and only if $(D_X\mathcal{E}^\bullet)[m]$ satisfies $\mathrm{AX}_\mathrm{q}(d)$ and $\mathrm{AX}_\mathrm{q}(c)$ respectively. Here q is the complementary perversity to p. In fact, if X is orientable then

$$\mathcal{E}^\bullet \text{ satisfies } \mathrm{AX}_\mathrm{p} \iff (D_X\mathcal{E}^\bullet)[m] \text{ satisfies } \mathrm{AX}_\mathrm{q}.$$

It follows from Theorem 7.2.5 that we have a canonical isomorphism

$$\mathcal{I}^q\mathcal{S}_X^\bullet \cong (D_X\mathcal{I}^p\mathcal{S}_X^\bullet)\,[m] \tag{7.3}$$

which, upon taking compactly supported hypercohomology groups, induces isomorphisms

$$I^qH_i(X) \cong I^pH^{cl}_{m-i}(X)^\vee.$$

These give rise to the non-degenerate pairing in Theorem 5.1.1. Hence, generalised Poincaré duality follows from the fact that Verdier duality interchanges the axioms AX_p with AX_q.

7.5   Further reading

The proofs of the assertions we have made in this chapter can be found in Goresky and MacPherson [70]. This paper also includes a proof of the Lefschetz hyperplane theorem for the intersection homology of a singular projective variety. Gelfand and Manin [63] contains a concise treatment, placed in a much wider context, which is extremely useful as an overview and reference.

A detailed account of the sheaf-theoretic machinery, in particular of the constructible derived category of sheaves on a stratified space, can be found in Borel *et al.* [24]. Another self-contained treatment can be found in the compendious Kashiwara and Schapira [97], which concentrates on the case when the underlying space is a Whitney stratified manifold. In this situation the theory can be greatly refined by considering the micro-local geometry, i.e. the geometry of the cotangent bundle. Dimca [56] provides a condensed and digestible exposition of this theory (with some proofs and technicalities omitted) and includes a wealth of examples and applications, mostly for complex analytic or algebraic varieties. See also Schürmann [159].

Beilinson, Bernstein and Deligne's seminal paper [13] puts the constructions of this chapter in the wider context of perverse sheaves (see the next chapter) and also explains how they can be applied to algebraic varieties over finite fields (see Chapter 10). Brylinski [36] provides a brief survey.

Chapter 8

Perverse sheaves

In this chapter X will be a $2r$-dimensional topological pseudomanifold and $\mathbb{S}$ will be a stratification of X with no odd-dimensional strata. The typical examples we have in mind are complex analytic or algebraic varieties Whitney stratified by subvarieties.

8.1 Perverse sheaves

Local systems (locally constant sheaves, representations of the fundamental group... see §4.9) are fundamental tools in geometry and topology. If one is working with stratified spaces then it is natural to consider a 'stratified local system', i.e. a sheaf which is locally constant on each stratum. These are precisely the constructible sheaves introduced in Definition 7.2.1. Thinking of a constructible sheaf as a complex concentrated in degree zero we see that the $\mathbb{S}$-constructible sheaves form an Abelian subcategory of the cohomologically constructible derived category $D^b_{\mathbb{S}}(X)$. At the end of the last chapter we introduced Verdier duality, a contravariant functor D_X from $D^b_{\mathbb{S}}(X)$ to itself which generalises the usual duality of vector spaces. It is natural to ask whether the Verdier dual of a constructible sheaf is again a constructible sheaf, and perhaps disappointing to discover that the answer is no. However, Beilinson, Bernstein and Deligne showed in [13] that there is another Abelian subcategory of $D^b_{\mathbb{S}}(X)$ which *is* preserved by Verdier duality. Furthermore this subcategory is intimately related to intersection homology, and arises naturally in a wide range of contexts in topology, analysis and algebra. The objects of this subcategory are the perverse sheaves (although we will see that they are not sheaves at all but rather complexes of sheaves).

Definition 8.1.1 (Beilinson, Bernstein and Deligne, [13, §2]). A complex of sheaves $\mathcal{E}^\bullet \in D^b_{\mathbb{S}}(X)$ is a **perverse sheaf** if for each stratum S and point $x \in S$

$$H^i(j_x^* \mathcal{E}^\bullet) = 0 \text{ for } i > -\dim S/2 \quad \text{and} \quad H^i(j_x^! \mathcal{E}^\bullet) = 0 \text{ for } i < \dim S/2. \quad (8.1)$$

Alternatively we can reformulate the conditions independently of the stratification: $\mathcal{E}^\bullet \in D^b_\mathbb{S}(X)$ is a perverse sheaf if

$$\dim\{H^{-i}(j_x^* \mathcal{E}^\bullet) \neq 0\} \leq 2i \quad \text{and} \quad \dim\{H^i(j_x^! \mathcal{E}^\bullet) \neq 0\} \leq 2i. \qquad (8.2)$$

The full subcategory of perverse sheaves in $D^b_\mathbb{S}(X)$ will be denoted $\mathbb{P}_\mathbb{S}(X)$. (We have formulated the conditions in terms of the stalks and costalks as in Kashiwara and Schapira [97, §10.3]. In Beilinson, Bernstein and Deligne [13] they are written in terms of j_S^* and $j_S^!$ where j_S is the inclusion of the stratum.)

Example 8.1.2. If X is a manifold trivially stratified with only one stratum then a perverse sheaf is a local system placed in degree $-\dim X/2$, i.e. a complex $\mathcal{L}[-\dim X/2]$ where $\mathcal{L}$ is a local system on X.

The conditions (8.1) and (8.2) closely resemble the axioms for the intersection homology sheaf. Let m be the lower middle perversity so that $m(2k) = k - 1$. Recalling that X and all strata of X are even-dimensional, axioms (c) and (d) on page 109 become

$$\begin{aligned} H^i(j_x^* \mathcal{E}^\bullet) &= 0 \quad \text{for } i > (-\dim S - \dim X)/2 \\ H^i(j_x^! \mathcal{E}^\bullet) &= 0 \quad \text{for } i < (\dim S - \dim X)/2 \end{aligned} \qquad (8.3)$$

respectively, *for x in a stratum S of codimension > 0.* Thus the precise relationship with the conditions for a complex to be a perverse sheaf is given by

Lemma 8.1.3. *A complex of sheaves $\mathcal{E}^\bullet \in D^b_\mathbb{S}(X)$ is a perverse sheaf if and only if the shift $\mathcal{E}^\bullet[r]$ (where $\dim X = 2r$) satisfies axioms (c) and (d) for the lower middle perversity — that is (8.3) — with the inequalities replaced by weak inequalities and without the restriction that $\operatorname{codim} S > 0$. In particular $\mathcal{IS}^\bullet_X[-r]$ is a perverse sheaf.*

The only point with which we need to take care is that $\mathcal{IS}^\bullet_X[-r]$ satisfies the conditions (8.1) for the codimension 0 strata of X. This follows since $\mathcal{IS}^\bullet_X$ satisfies (b) on page 109.

More generally, if S is a stratum of X then its closure $\overline{S} = Y$ is a closed union of strata of X and is naturally a stratified pseudomanifold with only even dimensional strata. Let $\mathcal{L}$ be a local system on S. Then we can define the intersection homology sheaf $\mathcal{IS}^\bullet_{Y,\mathcal{L}}$ on Y with coefficients in the local system $\mathcal{L}$. Its extension by zero (which in a mild abuse of notation we denote in the same way) is a cohomologically $\mathbb{S}$-constructible complex on X and we can verify that

$$\mathcal{IS}^\bullet_{Y,\mathcal{L}}[-s]$$

is a perverse sheaf, where $\dim S = 2s$. We can recover the stratum S and local system $\mathcal{L}$ from this perverse sheaf so we have a plentiful supply of examples of perverse sheaves, one for each local system on a stratum of X.

The perverse sheaves have many good properties, prominent amongst which are the following two theorems.

Theorem 8.1.4 (Beilinson, Bernstein and Deligne [13, §2]). *The category* $\mathbb{P}_{\mathbb{S}}(X)$ *of perverse sheaves is Abelian. A sequence*

$$0 \to \mathcal{E}^{\bullet} \to \mathcal{F}^{\bullet} \to \mathcal{G}^{\bullet} \to 0$$

of perverse sheaves is a short exact sequence if and only if there is a map $\mathcal{G}^{\bullet} \to \mathcal{E}^{\bullet}[1]$ *such that*

$$\mathcal{E}^{\bullet} \to \mathcal{F}^{\bullet} \to \mathcal{G}^{\bullet} \to \mathcal{E}^{\bullet}[1]$$

is a distinguished triangle in $D_{\mathbb{S}}^{b}(X)$.

The proof involves a general theory of Abelian subcategories of derived categories, the theory of t-structures, developed in Beilinson, Bernstein and Deligne [13] (see also Dimca [56, §5.1], Kashiwara and Schapira [97, §10.1] and Gelfand and Manin [63, Chapter 5 §3]).

Theorem 8.1.5 (Beilinson, Bernstein and Deligne, [13, §4]). *The Verdier dual* D_X *preserves the perverse sheaves; in fact it restricts to an exact contravariant functor from* $\mathbb{P}_{\mathbb{S}}(X)$ *to itself.*

The fact that it preserves the perverse sheaves follows from the discussion on page 115 which shows that the two conditions in (8.1) are dual to one another in the sense that if $\mathcal{E}^{\bullet}$ satisfies the first then $D_X \mathcal{E}^{\bullet}$ satisfies the second and vice-versa. The fact that D_X induces an *exact* functor on the Abelian category of perverse sheaves follows from the general theory of t-structures.

Example 8.1.6. Recall that a local system $\mathcal{L}$ on a space X is given by data consisting of a vector space $\mathcal{L}_x$ for each $x \in X$ and an isomorphism

$$\varphi^* \colon \mathcal{L}_{\varphi(0)} \to \mathcal{L}_{\varphi(1)}$$

for each continuous path $\varphi \colon [0,1] \to X$. There is a **dual local system** $\mathcal{L}^{\vee}$ with $\mathcal{L}_x^{\vee} = \mathrm{Hom}(\mathcal{L}_x, \mathbb{F})$ and with isomorphism

$$(\overline{\varphi}^*)^{\vee} \colon \mathcal{L}_{\varphi(0)}^{\vee} \to \mathcal{L}_{\varphi(1)}^{\vee}$$

corresponding to the path φ, where $\overline{\varphi}$ is the same path but traversed in the opposite direction. If X is a d-manifold then a direct computation shows that the Verdier dual of $\mathcal{L}$ (which we now think of as a locally constant sheaf) is

$$D_X \mathcal{L} = \mathcal{L}^{\vee}[d].$$

Using this computation (and the facts about the Verdier dual listed in Theorem 7.4.1) we can compute

$$D_X \left(\mathcal{IS}_{Y,\mathcal{L}}^{\bullet}[-s] \right) \cong \left(D_Y \mathcal{IS}_{Y,\mathcal{L}}^{\bullet} \right) [s] \cong \mathcal{IS}_{Y,\mathcal{L}^{\vee}}^{\bullet}[-s].$$

where, as before, Y is the closure of a $2s$-dimensional stratum S of X and $\mathcal{L}$ is a local system on S. This verifies the first part of Theorem 8.1.5 for the examples of perverse sheaves which we have met so far.

We saw above that the intersection homology sheaves on the closures of strata with coefficients in a local system are perverse sheaves. In fact every perverse sheaf is 'built up' from a finite number of these in a sense which we now explain.

Definition 8.1.7. An object a in an Abelian category A is **simple** if there are no non-trivial short exact sequences

$$0 \to p \to a \to q \to 0$$

in A (i.e. either p or q is zero).

Theorem 8.1.8 (cf. Beilinson, Bernstein and Deligne [13, Thm. 4.3.1]). *The category $\mathbb{P}_{\mathbb{S}}(X)$ of perverse sheaves is Artinian. In other words, every perverse sheaf $\mathcal{E}^{\bullet}$ has a finite length composition series*

$$0 \hookrightarrow \mathcal{E}_0^{\bullet} \hookrightarrow \mathcal{E}_1^{\bullet} \hookrightarrow \cdots \hookrightarrow \mathcal{E}_n^{\bullet} = \mathcal{E}^{\bullet}$$

for which the quotients $\mathcal{E}_i^{\bullet}/\mathcal{E}_{i-1}^{\bullet}$ are simple perverse sheaves, and the maximal length of any such composition series for $\mathcal{E}^{\bullet}$ is finite. (The quotients make sense because the category of perverse sheaves is Abelian.)

Furthermore, any simple perverse sheaf in $\mathbb{P}_{\mathbb{S}}(X)$ has the form $\mathcal{IS}_{Y,\mathcal{L}}^{\bullet}[-s]$ where Y is the closure of a connected stratum S of dimension $2s$ and $\mathcal{L}$ is an irreducible local system on S. (A local system is irreducible if the corresponding representation of the fundamental group is irreducible.)

In outline the argument goes as follows. If we have a (strict) descending chain

$$\mathcal{A}_0^{\bullet} \supset \mathcal{A}_1^{\bullet} \supset \ldots$$

of perverse sheaves then we can show that, for sufficiently large i, the perverse sheaf $\mathcal{A}_i^{\bullet}$ is supported (in the sense that its stalks vanish elsewhere) on a subset of strictly higher codimension than $\mathcal{A}_0^{\bullet}$. Because X is finite-dimensional we cannot increase the codimension indefinitely and so any descending chain of perverse sheaves must terminate, i.e. $\mathbb{P}_{\mathbb{S}}(X)$ is Noetherian. Since Verdier duality preserves the perverse sheaves we obtain the dual statement for free: ascending chains must also terminate, i.e. $\mathbb{P}_{\mathbb{S}}(X)$ is Artinian.

It is not hard to see that a simple perverse sheaf $\mathcal{E}^{\bullet}$ must be supported on the closure Y of a connected stratum S. The perverse conditions ensure that its restriction to S must be a local system $\mathcal{L}$, shifted by half the dimension of S. Suppose T is a stratum of minimal codimension in the closure of S. Then the following are equivalent:

- the restriction of $\mathcal{E}^{\bullet}$ to $S \cup T$ has no subobject or quotient perverse sheaves which are supported on T;

- $\mathcal{E}^{\bullet}$ obeys the strong vanishing conditions

$$H^i(j_x^* \mathcal{E}^{\bullet}) = 0 \quad \text{for } i \geq (-\dim T - \dim X)/2$$
$$H^i(j_x^! \mathcal{E}^{\bullet}) = 0 \quad \text{for } i \leq (\dim T - \dim X)/2$$

for $x \in T$.

By induction, adding a stratum at a time, we find that $\mathcal{E}^\bullet$ obeys the same axioms as, and therefore is isomorphic to, the shifted intersection homology complex

$$\mathcal{I}S^\bullet_{Y,\mathcal{L}}[-s]$$

and, furthermore, that the latter is simple if and only if $\mathcal{L}$ is irreducible.

Another way of expressing this is to say that Deligne's construction provides the *unique minimal extension* of a perverse sheaf over a stratum, in the sense that there are no subobject or quotient perverse sheaves of this extension which are supported on the newly-added stratum. In particular, the (shifted) intersection homology complex $\mathcal{I}S^\bullet_X[-\dim X]$ is the unique minimal extension of the constant sheaf on the non-singular part of X.

8.2 Perverse sheaves on varieties

If X is a quasi-projective complex algebraic variety then we can define a category $\mathbb{P}_{alg-c}(X)$ of (algebraic) perverse sheaves on X without choosing a stratification. An (algebraic) perverse sheaf on X is a complex in $D^b_{alg-c}(X)$ which satisfies the stratification-independent conditions (8.2), i.e.

$$\dim\{H^{-i}(j_x^*\mathcal{E}^\bullet) \neq 0\} \leq 2i \quad \text{and} \quad \dim\{H^i(j_x^!\mathcal{E}^\bullet) \neq 0\} \leq 2i.$$

Equivalently, it is a bounded complex $\mathcal{E}^\bullet$ such that there exists a Whitney stratification by algebraic subvarieties with respect to which $\mathcal{E}^\bullet$ is cohomologically constructible and satisifies the conditions (8.1).

There is an entirely analogous construction of (analytic) perverse sheaves $\mathbb{P}_{an-c}(X)$ on a complex analytic variety.

Both $\mathbb{P}_{alg-c}(X)$ and $\mathbb{P}_{an-c}(X)$ are Artinian Abelian categories, preserved by Verdier duality. The simple objects are once again intersection homology sheaves; more precisely they are the objects

$$\mathcal{I}S^\bullet_{Y,\mathcal{L}}[-s]$$

where $\mathcal{L}$ is an irreducible local system on the non-singular part of an irreducible subvariety Y of X. Moreover the perverse sheaves

$$\mathcal{I}S^\bullet_{Y,\mathcal{L}}[-s] \quad \text{and} \quad \mathcal{I}S^\bullet_{Y',\mathcal{L}'}[-s]$$

are isomorphic if and only if $Y \cap Y'$ is dense in Y and in Y' and

$$\mathcal{L}|_{Y\cap Y'} \cong \mathcal{L}'|_{Y\cap Y'},$$

see Beilinson, Bernstein and Deligne [13, 4.3.1].

8.3   Nearby and vanishing cycles

Perverse sheaves arise naturally in the study of singular fibres of complex polynomials. In this section we give the briefest introduction to this subject.

Suppose $f : \mathbb{C}^n \to \mathbb{C}$ is a polynomial. We will write $X = \mathbb{C}^n$ and $X_t = f^{-1}(t)$ for the fibre of X over the point $t \in \mathbb{C}$. There is a diagram of analytic varieties

$$
\begin{array}{ccccccc}
X_0 & \overset{\imath}{\hookrightarrow} & X & \overset{\jmath}{\longleftarrow} & X - X_0 & \overset{\pi}{\longleftarrow} & \widetilde{X - X_0} \\
\downarrow f & & \downarrow f & & \downarrow f & & \downarrow \\
\{0\} & \hookrightarrow & \mathbb{C} & \longleftarrow & \mathbb{C} - \{0\} & \underset{\pi}{\longleftarrow} & \widetilde{\mathbb{C} - \{0\}}
\end{array}
$$

where $\widetilde{\mathbb{C} - \{0\}}$ is the universal cover of $\mathbb{C} - \{0\}$ and $\widetilde{X - X_0}$ is the pullback of $X - X_0$ along the covering map

$$
\pi : \widetilde{\mathbb{C} - \{0\}} \to \mathbb{C} - \{0\},
$$

i.e. as a set it is $\{(x, t) \in (X - X_0) \times (\widetilde{\mathbb{C} - \{0\}}) \mid f(x) = \pi(t)\}$.

Definition 8.3.1. The **nearby cycles functor**

$$
\Psi_f : D^b_{an-c}(X) \to D^b_{an-c}(X_0)
$$

is given by $\Psi_f \mathcal{E}^\bullet = \imath^* Rp_* p^* \mathcal{E}^\bullet$ where $p = \jmath\pi : \widetilde{X - X_0} \to X$.

Recall that the pullback p^* of sheaves is left adjoint to the pushforward p_*. An important consequence is that there is a natural map

$$
\mathcal{E}^\bullet \to Rp_* p^* \mathcal{E}^\bullet,
$$

see, for example, Iversen [84, p98]. Restricting to the fibre X_0 by applying $\imath^*$ we have the **comparison map**

$$
\imath^* \mathcal{E}^\bullet \to \imath^* Rp_* p^* \mathcal{E}^\bullet = \Psi_f \mathcal{E}^\bullet.
$$

Definition 8.3.2. The **vanishing cycles functor**

$$
\Phi_f : D^b_{an-c}(X) \to D^b_{an-c}(X_0)
$$

is the mapping cone (see Definition 3.7.4) on the comparison map. Thus there is always a distinguished triangle

$$
\imath^* \mathcal{E}^\bullet \to \Psi_f \mathcal{E}^\bullet \to \Phi_f \mathcal{E}^\bullet \to \imath^* \mathcal{E}^\bullet[1]. \tag{8.4}
$$

In order to understand the geometric significance of the nearby and vanishing cycles functors we consider the simplest case; take $\mathcal{E}^\bullet = \mathbb{F}_X$ to be the constant sheaf on X with coefficients in the field $\mathbb{F}$.

Suppose that U is a neighbourhood of $0 \in \mathbb{C}$ such that every $t \in \mathbb{C} - \{0\}$ is a regular value of f. Then for $t \in U - \{0\}$ the fibres X_t are all homeomorphic and the restriction of f to $f^{-1}(U - \{0\})$ will be a locally trivial fibration called the Milnor fibration of f at 0 — see Milnor [134]. The pullback of this fibration to the universal cover is trivial and a fibre of it, i.e. a fibre of

$$\widetilde{X - X_0} \to \widetilde{\mathbb{C} - \{0\}}$$

for sufficiently small t, can be thought of as a 'generic fibre' of f near 0. Readers familiar with sheaf theory will not find it hard to prove the following proposition.

Proposition 8.3.3 (Dimca [56, Prop. 4.2.2]). *For any $0 \neq t \in U$ an identification of X_t with this generic fibre gives an isomorphism*

$$H^i(X_0; \Psi_f \mathbb{F}_X) \cong H^i(X_t). \tag{8.5}$$

Indeed, if $\jmath_x : x \hookrightarrow X_0$ is the inclusion of a point, then for sufficiently small t we have isomorphisms

$$H^i(\jmath_x^* \Psi_f \mathbb{F}_X) \cong H^i(X_t \cap B_x) \tag{8.6}$$

*where B_x is a small ball about x in $\mathbb{C}^n$ and $X_t \cap B_x$ is the **local Milnor fibre** at x. This should explain the term 'nearby cycles'.*

The long exact sequence of cohomology groups arising from the distinguished triangle (8.4) is

$$\cdots \to H^i(X_0) \to H^i(X_t) \to H^i(X_0; \Psi_f \mathbb{F}_X) \to H^{i+1}(X_0) \to \cdots$$

Geometrically it is simpler to think in terms of the dual sequence

$$\cdots \to H_{i+1}(X_0) \to H^i(X_0; \Psi_f \mathbb{F}_X)^\vee \to H_i(X_t) \to H_i(X_0) \to \cdots$$

The (dual of the) comparison map $H_i(X_t) \to H_i(X_0)$ arises from a continuous map $X_t \to X_0$ called the **specialisation map** (Goresky and MacPherson [71, §6]). Typically, specialisation involves collapsing certain cycles in the fibre X_t and the homology of these 'vanishing cycles' is given by the image of the second term in $H_*(X_t)$.

Theorem 8.3.4 (Goresky and MacPherson [71, §6], Brylinski [37] and Kashiwara and Schapira [97, 10.3.13]). *The right shifts $\Psi_f[-1]$ and $\Phi_f[-1]$ of the nearby and vanishing cycle functors preserve perverse sheaves, i.e. they restrict to functors*

$$\mathbb{P}_{an-c}(X) \to \mathbb{P}_{an-c}(X_0).$$

Example 8.3.5. Take $X = \mathbb{C}^2$ and $f : (x, y) \mapsto x^3 + y^3 - xy$. The special fibre X_0 is the affine part of the plane projective curve $\{(x : y : z) \in \mathbb{CP}^2 \mid x^3 + y^3 = xyz\}$ and the general fibre X_t for $t \neq 0$ is its smoothing (see Figure 8.1).

Figure 8.1: Topological pictures of $x^3 + y^3 - xy = 0$ (left) and its smoothing $x^3 + y^3 - xy = t \neq 0$ (right). The three dots represent three punctures; the corresponding projective curves in $\mathbb{CP}^2$ are obtained by filling in these points.

Figure 8.2: The local Milnor fibre of $f(x,y) = x^3 + y^3 - xy$ at the singular point $(0,0)$ is pictured above — the waist collapses to a point in the singular fibre. At all non-singular points the local Milnor fibre is contractible.

We can find the local Milnor fibres at points $(x,y) \in X_0$ (see Figure 8.2) and hence compute

$$H^i(j_x^* \Psi_f \mathbb{F}_X) \cong \begin{cases} \mathbb{F} & i = 0 \\ \mathbb{F} & i = 1, (x,y) = (0,0) \\ 0 & \text{otherwise.} \end{cases}$$

Similarly we can compute the cohomology of the costalks. Note that the nearby cycles form a perverse sheaf but *not* an intersection homology sheaf, i.e. not one of the form $\mathcal{IS}_{Y,\mathcal{L}}^\bullet$ for some subvariety Y and local system $\mathcal{L}$.

We can also compute that the vanishing cycles are supported at the singular point $(0,0)$ with

$$H^i(j_x^* \Phi_f \mathbb{F}_X) \cong \begin{cases} \mathbb{F} & i = 1, x = (0,0) \\ 0 & \text{otherwise.} \end{cases}$$

This corresponds to the 1-cycle, the waist of the Milnor fibre, which contracts to a point in the singular fibre.

A priori the singularity of X_0 at $(0,0)$ could arise in two ways; either two points in a nearby fibre could coalesce or a circle could contract to a point.

Since we know from Theorem 8.3.4 that the shifted vanishing cycles $\Phi_f \mathbb{F}_X$ must be a perverse sheaf we can deduce that it is the second possibility which occurs. (Of course we could also obtain this conclusion from global topological considerations.)

8.4   The decomposition theorem

One of the most important theorems about intersection homology is the decomposition theorem of Beilinson, Bernstein, Deligne and Gabber (see Beilinson, Bernstein and Deligne [13], Goresky and MacPherson [69] and MacPherson [124]).

Definition 8.4.1. Let $X \subseteq \mathbb{CP}^n$ and $Y \subseteq \mathbb{CP}^n$ be quasi-projective complex varieties. A **regular map**

$$\varphi \colon X \to Y$$

is a map such that for each $x = (x_0 : \ldots : x_n) \in X$ there exists homogeneous polynomials $f_0, \ldots, f_m$ in $n + 1$ variables, all of the same degree and not all vanishing on $(x_0, \ldots, x_n)$, such that

$$\varphi(y_0 : \ldots : y_n) = \big(f_0(y_0 : \ldots : y_n) : \ldots : f_m(y_0 : \ldots : y_n)\big)$$

for all $(y_0 : \ldots : y_n)$ in some neighbourhood U of x in X. The map φ is called **projective** if it can be factored as

$$X \xrightarrow{\psi} \mathbb{CP}^N \times Y \xrightarrow{\chi} Y$$

for some N, where ψ is an isomorphism (i.e. a regular map with a regular inverse) of X onto a closed subvariety of $\mathbb{CP}^N \times Y$ and χ is the projection of $\mathbb{CP}^N \times Y$ onto Y. Note that every fibre $\varphi^{-1}(y)$ of φ is then a quasi-projective subvariety of $\mathbb{CP}^N$.

Theorem 8.4.2 (Decomposition theorem for intersection homology). *Let $\varphi \colon X \to Y$ be a projective map between complex quasi-projective varieties. Then there exist closed subvarieties V_α of Y, local systems L_α on the nonsingular parts $(V_\alpha)_{\mathrm{nonsing}}$ of V_α and integers ℓ_α such that*

$$IH_j(X) \cong \bigoplus_\alpha IH_{j-\ell_\alpha}(V_\alpha, L_\alpha) \tag{8.7}$$

for all $j \geq 0$.

This decomposition is a consequence of a more powerful result about the behaviour of perverse sheaves with respect to projective maps.

Theorem 8.4.3 (Decomposition theorem for perverse sheaves). *Suppose $\varphi \colon X \to Y$ is a projective map of quasi-projective varieties and $\mathcal{E}^\bullet \in \mathbb{P}_{alg-c}(X)$*

is a perverse sheaf of 'geometric origin' on X. Then the pushforward $R\varphi_*\mathcal{E}^\bullet$ is isomorphic to a direct sum of (shifted) simple perverse sheaves.

In particular, since we can characterise the simple perverse sheaves on Y as intersection homology complexes (Theorem 8.1.8), there exist closed subvarieties V_α of Y, local systems L_α on the non-singular parts $(V_\alpha)_{\mathrm{nonsing}}$ of V_α and integers ℓ_α such that there is an isomorphism

$$R\varphi_*\mathcal{E}^\bullet \cong \bigoplus_\alpha \mathcal{IS}^\bullet_{V_\alpha, L_\alpha}[\ell_\alpha]$$

in the algebraically constructible bounded derived category $D^b_{alg-c}(Y)$.

We refer the reader to Beilinson, Bernstein and Deligne [13] for the definition of a perverse sheaf of 'geometric origin'. However, we note that the intersection homology complex $\mathcal{IS}^\bullet_X$ is always of geometric origin. (In fact it is conjectured that the assumption that $\mathcal{E}^\bullet$ is of geometric origin is unnecessary, see Drinfeld [58].)

The proof of Theorem 8.4.3 in [13, §6] involves relating the intersection homology of a variety to the ℓ-adic intersection cohomology (see Chapter 10) of varieties over fields of non-zero characteristic. Saito [151] gives an alternative proof, based on the relation of perverse sheaves to D-modules (see Chapter 11) and the theory of mixed Hodge modules. An overview and further references can be found in Saito [150].

In order to understand the connection between these two results we need to know that for any bounded complex $\mathcal{F}^\bullet$ on a topological space A and continuous map $f : A \to B$ we have isomorphisms

$$H^*(A; \mathcal{F}^\bullet) \cong H^*(B; Rf_*\mathcal{F}^\bullet) \quad \text{and} \quad H^*_c(A; \mathcal{F}^\bullet) \cong H^*_c(B; Rf_!\mathcal{F}^\bullet) \qquad (8.8)$$

of hypercohomology groups, see e.g. Gelfand and Manin [63, Ch. 4 §5]. Furthermore if f is a proper map then $Rf_* \cong Rf_!$. To obtain (8.7) from Theorem 8.4.3 we put $\mathcal{E}^\bullet = \mathcal{IS}^\bullet_X$ and note that, since projective maps are proper,

$$IH_i(X) \cong H^{-i}_c(X; \mathcal{IS}^\bullet_X) \cong H^{-i}_c(Y; R\varphi_*\mathcal{IS}^\bullet_X).$$

Remark 8.4.4. The decomposition theorem is an algebro-geometric result; it does not hold for more general stratified spaces and maps. However, Cappell and Shaneson [40] prove a weaker, but still very useful, result for any proper stratified map $f : X \to Y$ between Whitney stratified spaces with only even codimension strata. To do so they define an equivalence relation, which they term cobordism, on self-dual complexes of sheaves on such a space. (A **self-dual** complex of sheaves on X is a bounded, constructible complex $\mathcal{F}^\bullet$ together with an isomorphism

$$\mathcal{F}^\bullet \cong D_X\mathcal{F}^\bullet[\dim X]$$

to the shifted Verdier dual. The most important example is the intersection homology complex $\mathcal{IS}^\bullet_X$ equipped with the Poincaré duality isomorphism.)

They then prove that the pushforward $Rf_*\mathcal{F}^\bullet$ of any self-dual complex $\mathcal{F}^\bullet$ on X is cobordant to a direct sum

$$\bigoplus_S \mathcal{IS}^\bullet_{\overline{S}, \mathcal{L}_S}[\mathrm{codim}\, S/2] \qquad (8.9)$$

where the $\mathcal{L}_S$ are local systems, given explicitly in terms of the map f and the complex $\mathcal{F}^\bullet$, on the strata S of Y. The self-dual structure on this sum arises from non-degenerate bilinear forms on the local systems and Poincaré duality isomorphisms for the intersection homology complexes. Youssin [180] gives a proof which emphasizes that this result is a formal consequence of the structure of the constructible derived category of Y and its Abelian subcategory of perverse sheaves.

Since $\mathcal{IS}^\bullet_X$ is a self-dual complex on X we obtain an expression (8.9) for $Rf_*\mathcal{IS}^\bullet_X$. When f is a projective map of complex algebraic varieties then the terms in this expression can be related to those that appear in the decomposition theorem.

Cobordant self-dual complexes need not have the same hypercohomology so Cappell and Shaneson's result cannot be used to compute intersection homology groups. However, important invariants of self-dual complexes are preserved under cobordism, in particular the result has applications to the signatures and L-classes of stratified spaces (see Cappell and Shaneson [40] and Banagl, Cappell and Shaneson [6]).

Two special cases of the decomposition theorem are very important:

Proposition 8.4.5. *1. Suppose that $\varphi\colon X \to Y$ is a resolution of singularities of Y. That is, X is non-singular and φ is a surjective projective map which restricts to an isomorphism from a dense open subset of X to the non-singular part Y_{nonsing} of Y. Then there is a unique α, say α_0, such that $V_{\alpha_0} = Y$, and moreover $\ell_{\alpha_0} = 0$ and L_{α_0} is the constant local system $\mathbb{F}$. Thus*

$$H_j(X) \cong IH_j(X) \cong IH_j(Y) \oplus \left(\bigoplus_{\alpha \neq \alpha_0} IH_{j-\ell_\alpha}(V_\alpha, L_\alpha) \right).$$

In other words the intersection homology of any quasi-projective variety Y is a direct summand of the ordinary homology of any resolution of singularities X of Y.

2. Suppose that $\varphi\colon X \to Y$ is a projective map which is topologically a fibration, with fibre V (a projective variety). That is, every $y \in Y$ has an open neighbourhood U in Y such that there is a homeomorphism

$$\varphi^{-1}(U) \to U \times V$$

*such that $\varphi\colon \varphi^{-1}(U) \to U$ corresponds to the projection onto U. In such a situation there is a **Leray spectral sequence** for computing* *the cohomology of X which has E_2 term given by*

$$E_2^{p,q} = H^p(Y; H^q(V)).$$

Here $H^q(V)$ denotes the local system L on Y with stalk

$$L_y = H^q(V_y)$$

where $V_y = \varphi^{-1}(y) \cong V$ for each $y \in Y$ (Bott and Tu [26, p169], Griffiths and Harris [77, p463]).

There is also a Leray spectral sequence for computing the intersection cohomology of X with E_2 term $IH^p((Y; IH^q(V))$. The decomposition theorem for φ is equivalent to the degeneration of this spectral sequence at the E_2 term. That is, it says that

$$IH^j(X) \cong \bigoplus_{p+q=j} IH^p(Y; IH^q(V))$$

or equivalently

$$IH_j(X) \cong \bigoplus_{p+q=j} IH_p(Y; IH_q(V)).$$

In special circumstances we can compute the intersection homology of a complex variety using a resolution of singularities. The simplest example is that of small resolutions.

Definition 8.4.6. A resolution of singularities $\varphi\colon X \to Y$ is called **small** if for every $r > 0$

$$\operatorname{codim}\left\{y \in Y \,\middle|\, \dim \varphi^{-1}(y) \geq r\right\} > 2r,$$

i.e. the fibres are small in the sense that their dimension is $< r$ except on a subset of codimension $> 2r$.

Theorem 8.4.7 (Goresky and MacPherson [70, §6.2]). *If $\varphi\colon X \to Y$ is a small resolution then*

$$IH_*(X) \cong IH_*(Y) \cong H^*(Y).$$

To prove this we consider the complex $R\varphi_* \mathcal{IS}_X^\bullet$ of sheaves on Y. On the one hand by (8.8) its compactly supported hypercohomology is $IH_*(X)$. On the other hand we can show that it satisfies the axioms (a)–(d) on page 109 which characterise $\mathcal{IS}_Y^\bullet$ up to isomorphism in $D^b(Y)$.

Remark 8.4.8. Every quasi-projective complex variety has a resolution of singularities but not every quasi-projective complex variety has a small resolution. Some varieties have more than one small resolution. One can show, for example, that intersection cohomology has no natural ring structure generalising the cup product on ordinary cohomology by exhibiting a variety with two small resolutions whose cohomology rings are not isomorphic.

Example 8.4.9 (Cheeger, Goresky and MacPherson [48, §5.2]). If $a \leq b$ are positive integers the **Grassmann variety**

$$\mathrm{Gr}(a, \mathbb{C}^b) = \{a\text{-dimensional subspaces of } \mathbb{C}^b\}$$

is a non-singular projective variety of dimension $a(b-a)$.

If M is a fixed subspace of $\mathbb{C}^b$ and $c \leq a$ is a positive integer then

$$S = \{V \in \mathrm{Gr}(a, \mathbb{C}^b) \mid \dim V \cap M \geq c\}$$

is a projective subvariety of $\mathrm{Gr}(a, \mathbb{C}^b)$ called a **single condition Schubert variety**. There is a resolution of singularities $\varphi \colon \tilde{S} \to S$ where

$$\tilde{S} = \{(V, W) \in \mathrm{Gr}(a, \mathbb{C}^b) \times \mathrm{Gr}(c, \mathbb{C}^b) \mid W \subseteq V \cap M\}$$

and $\varphi(V, W) = V$. It is an easy exercise to show that the resolution φ is a small resolution. Thus

$$IH_*(S) \cong H_*(\tilde{S}).$$

If we choose an isomorphism of M with $\mathbb{C}^d$ where $d = \dim M$ then we can define

$$\rho \colon \tilde{S} \to \mathrm{Gr}(c, \mathbb{C}^b) \colon (V, W) \mapsto W.$$

It is easy to check that ρ is a projective fibration with fibre $\mathrm{Gr}(a-c, \mathbb{C}^{b-c})$. Thus the decomposition theorem applied to ρ tells us that

$$H_j(\tilde{S}) \cong \bigoplus_{p+q=j} H_p\big(\mathrm{Gr}(c, \mathbb{C}^b)\big); H_q\big(\mathrm{Gr}(a-c, \mathbb{C}^{b-c})\big).$$

Grassmann varieties are simply connected. It follows that any local system over $\mathrm{Gr}(c, \mathbb{C}^b)$ is trivial. Hence we have

$$IH_j(S) \cong H_j(\tilde{S}) \cong \bigoplus_{p+q=j} H_p\big(\mathrm{Gr}(c, \mathbb{C}^b)\big) \otimes H_q\big(\mathrm{Gr}(a-c, \mathbb{C}^{b-c})\big).$$

The homology of Grassmann varieties is well known (see e.g. Griffiths and Harris [77, Ch. 1 §5]. The Betti numbers

$$B_j = \dim H_j\big(\mathrm{Gr}(a, \mathbb{C}^b)\big)$$

of $\mathrm{Gr}(a, \mathbb{C}^b)$ are given by the formula

$$\sum_{j \geq 0} B_j t^j = \frac{\prod_{a \leq i < b}(1 + t^2 + t^4 + \ldots + t^{2i})}{\prod_{1 \leq j < b-a}(1 + t^2 + t^4 + \ldots + t^{2j})}.$$

8.5   Further reading

The full theory of perverse sheaves on algebraic varieties is developed in Beilinson, Bernstein and Deligne [13]. Rietsch [146] is a nice introduction. For perverse sheaves on analytic varieties and a treatment of the micro-local geometry of perverse sheaves (an aspect of the subject we have ignored) see Kashiwara and Schapira [97] or, for a gentler introduction, Dimca [56]. The latter contains several applications including a full discussion of nearby and vanishing cycles.

Goresky and MacPherson [69] is a nice survey of applications of the decomposition theorem to the homology of complex algebraic varieties.

We defined perverse sheaves as an Abelian subcategory of the bounded derived category of sheaves but there are alternative approaches which avoid this formalism.

- MacPherson and Vilonen [126] provide an 'elementary' construction of perverse sheaves. The idea is to build up the category of perverse sheaves stratum-by-stratum by 'glueing' together Abelian categories; see also Beilinson [10, 11], Kashiwara [93], Verdier [174] and Gelfand and Manin [63, Ch. 7].

- The Riemann–Hilbert correspondence (see Chapter 11) describes the perverse sheaves on a complex algebraic variety as a natural category of D-modules. Very roughly, a perverse sheaf can be thought of as the solutions to a certain kind of system of differential equations.

- MacPherson [125] uses ideas from stratified Morse theory to define perverse sheaves in a very geometric way.

- In some special cases it is possible to give a quiver presentation of a category of perverse sheaves, i.e. a description in terms of linear algebra — see e.g. Gelfand, MacPherson and Vilonen [62], Braden and Grinberg [29] and Braden [28].

Perverse sheaves and intersection cohomology play an important rôle in representation theory. This is too large a topic (and the authors' ignorance too profound) to which to do justice in a short space, but here are three examples which serve as pointers.

1. Let G be a complex simple algebraic group and B a Borel subgroup. The flag variety G/B has a natural stratification by Schubert cells. The category of perverse sheaves on G/B is equivalent to a category whose simple objects are certain irreducible representations of the Lie algebra $\mathfrak{g}$ (Beilinson, Ginzburg and Soergel [14]). Perverse sheaves on flag varieties and the Kazhdan–Lusztig conjecture form the subject matter of Chapter 12.

2. Let k be an algebraic closure of a finite field $\mathbb{F}_p$. The general linear group $GL_n(\mathbb{F}_p)$ is the fixed point set of the Frobenius automorphism

$$F : GL_n(k) \rightarrow GL_n(k) : (g_{ij}) \mapsto (g_{ij}^p).$$

   The group $GL_n(k)$ acts on itself by conjugation and a **character sheaf** is an equivariant perverse sheaf on $GL_n(k)$ with respect to this action. The reason for the name is that we can associate a character of $GL_n(\mathbb{F}_p)$ to each character sheaf which is invariant under the Frobenius automorphism F. This sets up a bijection between irreducible characters of $GL_n(\mathbb{F}_p)$ and (isomorphism classes of) character sheaves invariant under F. In fact the theory of character sheaves applies much more widely than just to general linear groups. It has been extensively studied by Lusztig in a series of papers beginning with [119]. See [122] for a survey.

3. Let G be a complex reductive group and $\mathfrak{g}$ its Lie algebra. G acts on $\mathfrak{g}$ by the adjoint action. The most singular fibre of the associated quotient map $\mathfrak{g} \rightarrow \mathfrak{g}/G$ is called the **Springer fibre**. The nearby cycles of the constant sheaf at this fibre form a perverse sheaf which decomposes as a direct sum of simple perverse sheaves. The summands parameterise the irreducible representations of the Weyl group of G — see Borho and MacPherson [25].

Further information on the applications of perverse sheaves in representation theory can be found in Lusztig's survey [121].

Chapter 9

The intersection cohomology of fans

In the first few sections of this chapter we present some aspects of the geometry of toric varieties. These form a rich set of examples of complex varieties. Their properties are closely related to combinatorics and consequently are often unusually tractable to computation. By and large we omit proofs; these can be found in Danilov [50] or in Fulton [61]. As the name might suggest, torus actions play a central rôle in the theory of toric varieties and so we will be interested in *equivariant* cohomology and intersection cohomology — §9.7 provides a brief introduction sufficient for our present purposes. In sections 9.8 and 9.9 we discuss the intersection cohomology of toric varieties, its generalisation to fan spaces and the relation of these with Stanley's conjectures [167] on the combinatorics of polytopes.

We begin with the notion of a **torus embedding**: this is an algebraic variety X which

1. contains an algebraic torus $T \cong (\mathbb{C}^*)^n$ as a dense open subset,

2. has an action $T \times X \to X$ of T which extends the standard left action of T on itself by mutiplication.

Although this definition has the advantage of brevity it tells us little about the structure of such varieties. In order to elucidate the latter, we will give a much lengthier definition of a **toric variety** and then indicate that it amounts to the same thing.

9.1   Affine toric varieties

Let L be an n-dimensional lattice, i.e. an additive Abelian group isomorphic to $\mathbb{Z}^n$, and let $V = L \times_{\mathbb{Z}} \mathbb{R}$ be the associated real vector space.

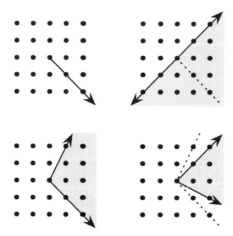

Figure 9.1: On the left, two examples of strongly convex rational cones in $\mathbb{Z}^2$; on the right, their duals (which are rational and convex but not necessarily strongly convex). We have implicitly identified the lattice with its dual using the standard basis for $\mathbb{Z}^2$.

Definition 9.1.1. A **convex cone** σ in V is a subset

$$\left\{ \sum_{i=1,\ldots,k} a_i v_i \mid a_i \in \mathbb{R}_{\geq 0}, v_i \in V, k \in \mathbb{N} \right\}.$$

The dimension of a cone is the dimension of the subspace spanned by the generating set $\{v_1, \ldots, v_k\}$. The **support** of a cone is the subset $|\sigma| = \{v \in \sigma\}$ of V. A convex cone is **strongly convex** if it does not contain any line through the origin.

A convex cone is **rational in** L if we can choose the generators $v_1, \ldots, v_k$ to lie in the lattice L. If the lattice is understood we will simply refer to this as a rational cone.

The **dual** of a cone σ is the subset

$$\sigma^\vee = \{l \in V^\vee \mid l(v) \geq 0 \ \forall v \in \sigma\}$$

of the dual vector space $V^\vee$. The dual of a cone is also a convex cone, and it is rational in the dual lattice

$$L^\vee = \mathrm{Hom}(L, \mathbb{Z}) \subset V^\vee$$

if σ is rational in L. (See Figure 9.1.)

Lemma 9.1.2 (Gordon's lemma). *If σ is a rational cone in L then the intersection*

$$M_\sigma = L^\vee \cap \sigma^\vee$$

is a finitely-generated monoid, or semigroup, under addition; that is, it is a set equipped with a binary operation which obeys the group axioms except for the existence of inverses. Furthermore, if σ is strongly convex then

$$M = M_\sigma + (-M_\sigma).$$

We will be interested almost exclusively in strongly convex cones and so henceforth whenever we use the term 'cone' without qualification we will mean 'strongly convex cone'.

Definition 9.1.3. If M is a monoid then the **free $\mathbb{C}$-algebra over** M is the algebra $\mathbb{C}[M]$ over $\mathbb{C}$ which as a vector space has basis $\{X_u \mid u \in M\}$ with multiplication given by

$$X_u X_v = X_{u+v}.$$

If M is finitely-generated as a monoid then $\mathbb{C}[M]$ is finitely-generated as an algebra.

The **affine toric variety** X_σ associated to a rational cone σ is defined as follows. We choose a set of generators $\{X_1, \ldots, X_r\}$ for $\mathbb{C}[M_\sigma]$ and this allows us to write

$$\mathbb{C}[M_\sigma] \cong \mathbb{C}[X_1, \ldots, X_r]/I$$

for some ideal I. The affine variety X_σ is defined to be the vanishing set of the polynomials in I. A different choice of generators leads to an isomorphic variety; X_σ is well-defined up to isomorphism.

Remark 9.1.4. A more sophisticated approach would be to define $X_\sigma = \mathrm{Spec}(\mathbb{C}[M_\sigma])$.

The elements of $\mathbb{C}[M_\sigma]$ are polynomial functions on the variety X_σ. A point in X_σ determines an evaluation map

$$\mathbb{C}[M_\sigma] \to \mathbb{C}$$

which is a surjective map of algebras. Conversely such a map uniquely determines a point of X_σ. This leads to an invariant description of the points of X_σ as maps $M_\sigma \to \mathbb{C}$ of monoids, where $\mathbb{C} = \mathbb{C}^* \cup \{0\}$ is considered as a monoid under multiplication. Explicitly, a map $x : M_\sigma \to \mathbb{C}$ gives a map

$$\mathbb{C}[M_\sigma] \to \mathbb{C} : X_u \mapsto x(u).$$

It is easy to check that $X_0 \mapsto x(0) = 1$ and

$$X_u X_v = X_{u+v} \mapsto x(u+v) = x(u)x(v)$$

so that this is a surjective map of algebras.

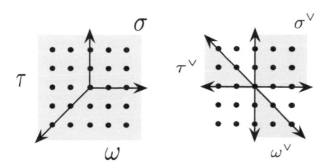

Figure 9.2: Three 2-dimensional cones in $\mathbb{R}^2$ and their duals. In the next section we will see that this is a picture of $\mathbb{CP}^2$.

Examples 9.1.5.   1. Consider the cone $\{0\}$ in $L \cong \mathbb{Z}^n$. The monoid $M_{\{0\}}$ is isomorphic to $(\mathbb{Z}_{\geq 0})^{2n}$. The corresponding algebra is generated by $2n$ elements $X_1, \ldots, X_n, Y_1, \ldots Y_n$ with relations $X_i Y_i = 1$. The corresponding affine variety is the algebraic torus $(\mathbb{C}^*)^n$ which we will denote by T_L.

2. Consider the three 2-dimensional cones in $\mathbb{Z}^2$ shown in Figure 9.2. The monoid M_σ is

$$\mathbb{Z}_{\geq 0} \cdot e_1 + \mathbb{Z}_{\geq 0} \cdot e_2,$$

where e_1 and e_2 are the standard basis of $L^\vee$, and the monoids M_τ and M_ω are

$$\mathbb{Z}_{\geq 0} \cdot (-e_1) + \mathbb{Z}_{\geq 0} \cdot (e_2 - e_1) \text{ and } \mathbb{Z}_{\geq 0} \cdot (e_1 - e_2) + \mathbb{Z}_{\geq 0} \cdot (-e_2)$$

respectively. The corresponding algebras are

$$\mathbb{C}[X, Y], \mathbb{C}[X^{-1}, X^{-1}Y] \text{ and } \mathbb{C}[XY^{-1}, Y^{-1}],$$

all of which are free algebras on two generators and so correspond to the variety $\mathbb{C}^2$. Check that the three 1-dimensional cones correspond respectively to the algebras

$$\mathbb{C}[X, X^{-1}, Y], \mathbb{C}[XY^{-1}, X^{-1}Y, X^{-1}Y^{-1}] \text{ and } \mathbb{C}[X, Y, Y^{-1}]$$

and that these algebras all give the variety $\mathbb{C} \times \mathbb{C}^*$.

A non-zero element $l \in V^\vee$ determines a hyperplane $l^\perp = \{l = 0\}$ in V. Such a hyperplane is said to **support** a cone σ if $l(v) \geq 0$ for all $v \in \sigma$. A **face** of σ is the intersection of σ with a supporting hyperplane. Any face of a cone is also a cone, and if τ is a face of σ we write $\tau < \sigma$.

Lemma 9.1.6 (See Fulton [61, §1.2]). *If $\tau = \sigma \cap l^{\perp}$ is a face of a rational cone σ in L then the monoid M_{τ} arises from M_{σ} by inverting the single element l, i.e.*

$$M_{\tau} = M_{\sigma} + \mathbb{Z}_{\geq 0}(-l).$$

An immediate consequence is that the algebra $\mathbb{C}[M_{\tau}]$ is obtained from $\mathbb{C}[M_{\sigma}]$ by localising at the element X_l, i.e.

$$\mathbb{C}[M_{\tau}] = \mathbb{C}[M_{\sigma}]_{(X_l)}.$$

It follows that the variety X_{τ} can be identified with the dense open subset

$$X_{\sigma} \cap \{X_l \neq 0\}$$

of X_{σ}. Since the cone $\{0\}$ is a face of every cone the torus $X_{\{0\}} \cong T_L \cong (\mathbb{C}^*)^n$ is a dense open subset of any X_{σ}.

9.2 Toric varieties from fans

Definition 9.2.1. A **fan** Φ in the vector space V is a finite set of cones in V such that

- if $\sigma \in \Phi$ then any face of σ is in Φ;

- the intersection of any two cones in Φ is also a cone in Φ.

The fan Φ is **rational** if each cone in Φ is rational. The **support** of the fan is the closed subset

$$|\Phi| = \cup_{\sigma \in \Phi} |\sigma|$$

and we say the fan is **complete** if the support is the whole of V.

For a subset S of cones we define $[S]$ to be the minimal subfan containing S (defined as the intersection of all subfans containing S). The **boundary** of S is the subset $\partial S = [S] - S$. The **star** of S is the subset $\text{Star}(S) = \{\tau \geq \sigma | \sigma \in S\}$.

Suppose Φ is a rational fan in V with respect to the lattice L. The (abstract) **toric variety** X_{Φ} **associated to** Φ is constructed as follows. For each cone σ we have a complex variety X_{σ}. The fan determines a 'glueing diagram'; for cones σ and τ in Φ the intersection is a face of both and we have open inclusions

$$X_{\sigma} \hookleftarrow X_{\sigma \cap \tau} \hookrightarrow X_{\tau}.$$

The variety X_{Φ} is obtained by identifying the image of $X_{\sigma \cap \tau}$ in X_{σ} and X_{τ} for each pair of cones σ and τ in Φ. The subsets $X_{\{0\}} \cong T_L$ of each X_{σ} are glued together and it follows that there is a dense open subset of X_{Φ} isomorphic to T_L.

Remark 9.2.2. We say *abstract* variety because the result of glueing, although locally affine, need not be quasi-projective, i.e. it need not arise as a subset of projective space cut out by polynomials. We will give a necessary and sufficient criterion for X_Φ to be a projective variety below (Proposition 9.4.1).

Example 9.2.3. Consider the fan depicted in Figure 9.2. We identified the affine toric varieties corresponding to the cones in the fan in Examples 9.1.5. Each 2-dimensional cone corresponds to a copy of $\mathbb{C}^2$ with coordinates X, Y for X_σ, coordinates $X^{-1}, X^{-1}Y$ for X_τ and coordinates $Y^{-1}, Y^{-1}X$ for X_ω. The resulting glued-up variety is $\mathbb{CP}^2$.

This correspondence between fans and varieties is very natural. For example if we have fans Φ and Φ' in lattices L and L' respectively then there is a product fan $\Phi \times \Phi'$ in the product lattice $L \times L'$ whose cones are of the form $\sigma \times \sigma'$. By definition the monoid

$$M_{\sigma \times \sigma'} = \{(l, l') \in L \times L' | (l, l') \cdot (v, v') \geq 0 \quad \forall (v, v') \in \sigma \times \sigma'\}.$$

Since $(l, l') \cdot (v, v') = l(v) + l'(v')$ and $0 \in \sigma, \sigma'$ we see that we must have $l \in M_\sigma$ and $l' \in M_{\sigma'}$, i.e.

$$M_{\sigma \times \sigma'} = M_\sigma \times M_{\sigma'}.$$

It follows that $\mathbb{C}[M_{\sigma \times \sigma'}] = \mathbb{C}[M_\sigma] \otimes \mathbb{C}[M_{\sigma'}]$ and so the corresponding affine toric variety is simply the product $X_\sigma \times X_{\sigma'}$. This is all compatible with glueing so we conclude that the product of fans corresponds to the product of toric varieties.

In the next section we will see that maps of fans correspond to maps of varieties.

9.3   Maps and torus actions

If τ is a face of a cone σ then we saw earlier that there is an open embedding

$$X_\tau \hookrightarrow X_\sigma.$$

This is a special case of

Proposition 9.3.1. *Suppose Φ and Φ' are rational fans with respect to lattices L and L' and that $\alpha : L \to L'$ is a map of lattices such that for each cone $\sigma \in \Phi$ the image $\alpha(\sigma)$ is contained in a cone of Φ'. Then there is an induced map*

$$\alpha_* : X_\Phi \to X_{\Phi'}.$$

Proof. If $\alpha(\sigma) \subset \sigma'$ then the dual map

$$\alpha^\vee : (L')^\vee \to L^\vee$$

takes $M_{\sigma'}$ to M_σ. This is because for $f \in M_{\sigma'}$ and $v \in \sigma$ we have

$$\alpha^\vee(f)(v) = f(\alpha(v)) \geq 0$$

since $\alpha(v) \in \sigma'$. Hence $\alpha^\vee$ induces a map of algebras $\mathbb{C}[M_{\sigma'}] \to \mathbb{C}[M_\sigma]$. Any such map induces a map $X_\sigma \to X_{\sigma'}$ (see, for example, Hartshorne [79, Ch. II, Prop. 2.3]). We can check that these maps are compatible and so patch together to give the desired map $X_\Phi \to X_{\Phi'}$.   □

Examples 9.3.2.   1. If we take α to be the identity on L and Φ to be a subfan of Φ' then the induced map $X_\Phi \to X_{\Phi'}$ is an open embedding. We have met this before in the case of a face of a cone.

2. Suppose we take Φ and Φ' to be the cone $\{0\}$. Then we obtain a map $T_L \to T_{L'}$ of tori, which is a map of groups (as well as of varieties). Embeddings of lattices correspond to embeddings of tori and quotients of lattices to quotients of tori.

We can also use this construction to see that all toric varieties carry torus actions. Given the lattice L there is a map $L \times L \to L$ given by the group operation of addition. The cones of the fan $\{0\} \times \Phi$ in $L \times L$ clearly map into the cones of Φ under this. Since the product of fans corresponds to the product of varieties and the cone $\{0\}$ in L corresponds to the torus T_L we obtain a map

$$T_L \times X_\Phi \to X_\Phi$$

It is easy to check that this defines an action of T_L on X_Φ.

Exercise 9.3.3. When $\Phi = \{0\}$ show that the resulting action is the standard (left) action of T_L on itself by multiplication.

These torus actions are natural in the sense that if $\alpha : L \to L'$ is a map of lattices inducing a map $\alpha_* : X_\Phi \to X_{\Phi'}$ as above then we also have a map $\alpha : T_L \to T_{L'}$ of tori and these maps satisfy

$$\alpha_*(t \cdot x) = \alpha_*(t) \cdot \alpha_*(x)$$

for $t \in T_L$ and $x \in X_\Phi$. In particular, by considering the open embedding $T_L \hookrightarrow X_\Phi$ arising from the inclusion $\{0\} \hookrightarrow \Phi$ of fans we see that the action of T_L on X_Φ extends the usual action of T_L on itself. Hence the toric variety X_Φ is a torus embedding. Indeed, although we will neither show nor use this fact, all normal, separable torus embeddings arise in this way as the toric variety associated to some rational fan (Oda, [139, Thm. 4.1]).

The description of the torus action on a toric variety above is rather abstract and it is useful to have a more down-to-earth description too. Suppose σ is an rational fan with respect to L. Recall that the points of the torus T_L and of X_σ correspond respectively to monoid maps $L^\vee \to \mathbb{C}$ and $M_\sigma \to \mathbb{C}$

where $\mathbb{C} = \mathbb{C}^* \cup \{0\}$ is made into a monoid under multiplication. Given elements $t \in T_L$ and $x \in X_\sigma$ we define $t \cdot x$ to be the point of X_σ corresponding to the map

$$M_\sigma \to \mathbb{C} : u \mapsto t(u)x(u).$$

We can verify that these actions on affine toric varieties are compatible with the glueing procedure and thus obtain an action of T_L on any toric variety. It is a good exercise in unwinding abstractions to show that this agrees with the earlier definition!

Examples 9.3.4.   1. When σ is the cone $\{0\}$ in an n-dimensional lattice L then points of the associated affine toric variety T_L correspond to monoid maps $L^\vee \to \mathbb{C} = \mathbb{C}^* \cup \{0\}$ and the action of the torus on itself defined above is given by pointwise multiplication of functions. This corresponds to the usual multiplication of functions in the algebra

$$\mathbb{C}[L^\vee] \cong \mathbb{C}[X_1, X_1^{-1}, \ldots, X_n, X_n^{-1}].$$

In terms of the toric variety T_L this is (of course) the standard left action of the torus on itself. In particular the action is free.

2. At the other extreme of dimension, if σ is an n-dimensional cone in V then the affine toric variety X_σ has a unique fixed point under the action of T_L. This is because a point $x \in X_\sigma$, which we think of as a map of monoids $M_\sigma \to \mathbb{C}$, is fixed if and only if

$$t(u)x(u) = x(u) \quad \forall u \in M_\sigma, t \in T_L \iff x(u) = 0 \quad \forall 0 \neq u \in M_\sigma.$$

Since x is a map of monoids (where $\mathbb{C}$ is a monoid under multiplication) we must have $x(0) = 1$ so that the map x is uniquely determined. We will refer to this distinguished fixed point as x_σ.

9.4   Projective toric varieties and convex polytopes

We will be particularly interested in toric varieties which are projective, i.e. which can be embedded as a subset of projective space cut out by homogeneous polynomials. If a toric variety is to be projective it must be compact (in the analytic topology). It is not hard to see that X_Φ is compact if and only the fan Φ is complete. However this is not sufficient to guarantee that X_Φ is projective.

Proposition 9.4.1 (See Fulton [61, §2.4]). *The toric variety X_Φ is projective if and only if the fan Φ is complete and there exists a continuous function $\ell : V \to \mathbb{R}$ such that*

- *ℓ is **cone-wise linear**, i.e. the restriction of ℓ to the support $|\sigma|$ of any cone is a linear function $v \mapsto \ell_\sigma(v)$ for some $\ell_\sigma \in L^\vee$;*

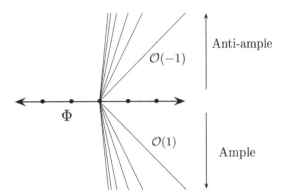

Figure 9.3: The group $\mathrm{Pic}(\mathbb{CP}^1)$ represented as (the graphs of) continuous cone-wise linear functions on the fan Φ — see Remark 9.4.2.

- ℓ is **strictly convex**, *i.e.* $\ell_\sigma(v) > \ell(v)$ *for* $v \in V - |\sigma|$.

We will call such a function **ample** *and will call a complete fan equipped with such an ample function a* **projective** *fan.*

Remark 9.4.2. For readers familiar with algebraic geometry the explanation of this result and terminology is as follows. There is an isomorphism between the group $\mathrm{Pic}(X_\Phi)$ of isomorphism classes of line bundles on X_Φ and the quotient of the group of cone-wise linear continuous functions on the support $|\Phi| = V$ by the subgroup of linear functions. Under this identification ample line bundles correspond to the classes of strictly convex cone-wise linear functions — see Fulton [61, §3.4].

As an example consider the toric variety $\mathbb{CP}^1$ corresponding to the unique complete fan Φ in dimension 1. On the one hand it is well-known that $\mathrm{Pic}(\mathbb{CP}^1) \cong \mathbb{Z}$ with the ample bundles corresponding to negative integers. On the other hand, a continuous cone-wise linear function on Φ corresponds to a pair of linear functions

$$l_- : \mathbb{Z}_{\leq 0} \to \mathbb{Z} \quad \text{and} \quad l_+ : \mathbb{Z}_{\geq 0} \to \mathbb{Z}$$

which agree at the origin. Each class in the quotient by the linear functions is represented by a unique continuous cone-wise linear function which vanishes on $\mathbb{Z}_{\leq 0}$ — see Figure 9.3. Such a representative is determined by its value at 1, i.e. by a single integer. The function is strictly convex if and only if this integer is negative.

The data of a complete fan and an ample function is equivalent to the data of a convex polytope. Thus there is also a correspondence between convex polytopes and projective toric varieties which we now describe.

A **convex polytope** P in $V^\vee$ is the convex hull of a finite set of points. It is **rational** if these points lie in the lattice $L^\vee$. We will always assume that

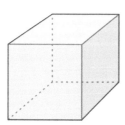 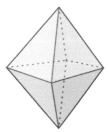

Figure 9.4: The polar polytope of the cube with vertices at $(\pm 1, \pm 1, \pm 1)$ is the octahedron with vertices at $(\pm 1, 0, 0), (0, \pm 1, 0), (0, 0, \pm 1)$.

the origin 0 is contained in the interior of P which, in particular, ensures that P is n-dimensional.

Given $v \in V$ and $d \in \mathbb{R}$ there is a hyperplane

$$H = \{l \in V^{\vee} | l(v) = d\}.$$

We say H **supports** P if $p(v) \geq 0$ for all $p \in P$. A **proper face** is the intersection of P with a supporting hyperplane. Note that we allow the empty set as a proper face.

A rational convex polytope P (with the origin in its interior) determines a rational fan Φ_P with one cone for each proper face. This fan is the fan over the **polar polytope** P^o which is the rational convex polytope in V given by

$$\{v \in V \mid p(v) \geq -1 \; \forall p \in P\}.$$

(See Figure 9.4.) We recall the following properties of polar polytopes:

1. if F is a face of P then $F^o = \{v \in P^o \mid f(v) = -1 \; \forall f \in F\}$ is a face of P^o with
$$\dim F + \dim F^o = \dim V - 1,$$

2. the mapping $F \mapsto F^o$ is an inclusion-reversing bijection between faces of P and of P^o,

3. the polytope P is the polar of P^o.

Let Φ_P be the fan over the polar polytope P^o. In other words Φ_P has one cone

$$\{\lambda v \mid v \in F^o, \lambda \in \mathbb{R}_{\geq 0}\}$$

for each face F of P (with the cone $\{0\}$ corresponding to the empty face). The fan Φ_P is complete because it is the fan over a polytope with 0 in the interior. Furthermore, there is an ample function

$$\ell_P : |\Phi_P| \to \mathbb{R}$$

given by $\ell_P(v) = \min p_i(v)$ where p_i ranges over the vertices of P. We denote the projective toric variety associated to the fan Φ_P by X_P.

Conversely, a rational fan Φ in V and an ample function $\ell : |\Phi| \to \mathbb{R}$ determine a convex polytope

$$P(\Phi) = \{f \in V^\vee \mid f(v) \ge \ell(v) \; \forall v \in |\Phi|\}$$

in the dual space $V^\vee$.

Exercise 9.4.3. Show that these constructions are inverse to one another, i.e. that $\Phi = \Phi_{P(\Phi)}$ and $\ell = \ell_{P(\Phi)}$.

To recapitulate: there are correspondences

$$
\begin{array}{rcl}
\text{toric variety} & \leftrightarrow & \text{rational fan} \\
\text{compact toric variety} & \leftrightarrow & \text{complete rational fan} \\
\text{projective toric variety with} & \leftrightarrow & \text{complete rational fan with} \\
\text{a chosen ample line bundle} & & \text{a chosen ample function} \\
& \leftrightarrow & \text{rational convex polytope}
\end{array}
$$

Remark 9.4.4. There are complete rational fans on which there are no ample functions (see Fulton [61, p71]). Hence there are compact toric varieties which are not projective and there are fans which do not arise as the fan over any convex polytope.

9.5 Stratifications of toric varieties

Suppose L is an n-dimensional lattice and Φ is a rational fan in L. The associated toric variety X_Φ has a decomposition into orbits under the action of the torus $T = T_L$. In this section we explain how these orbits can be explicitly described in terms of Φ.

Recall that X_Φ is covered by open subsets X_σ for $\sigma \in \Phi$. We consider the structure of the affine toric variety X_σ in more detail. The elements of $L \cap \sigma$ generate a sublattice of L with real span the subspace $\mathrm{Span}(\sigma)$ of V. Clearly σ and its faces form a rational fan in L_σ which we denote $\langle\sigma\rangle$. Write T_σ for the torus T_{L_σ}. It acts on the affine toric variety $X_{\langle\sigma\rangle}$ and, by Example 9.3.4, there is a unique fixed point x_σ.

The inclusion $L_\sigma \hookrightarrow L$ induces a map $f : X_{\langle\sigma\rangle} \to X_\sigma$ which is T_σ-equivariant (where T_σ acts on X_σ via its inclusion into T). In this situation we can define a T-equivariant map

$$T \times X_{\langle\sigma\rangle} \to X_\Phi : (t, x) \mapsto t \cdot f(x) \tag{9.1}$$

where T acts by left multiplication on the first factor on the left-hand side. For any $s \in T_\sigma$ we have

$$ts \cdot f(x) = t \cdot f(sx)$$

so that (9.1) factors through the quotient $T \times_{T_\sigma} X_{\langle \sigma \rangle}$ of $T \times X_{\langle \sigma \rangle}$ by the action $s \cdot (t, x) = (ts^{-1}, s \cdot x)$ of T_σ.

Proposition 9.5.1. *The induced map $T \times_{T_\sigma} X_{\langle \sigma \rangle} \to X_\sigma$ is a T-equivariant isomorphism. In particular, the image of*

$$T \times_{T_\sigma} \{x_\sigma\} \cong T/T_\sigma \cong (\mathbb{C}^*)^{n - \dim \sigma},$$

is an orbit of T in X_σ which we will denote by $\mathcal{O}_\sigma$. The complement of this orbit in X_σ is naturally identified with the toric variety $X_{\partial \sigma}$ associated with the boundary fan of σ.

Clearly we can also consider $\mathcal{O}_\sigma$ to be an orbit in X_Φ. Indeed every orbit is of this form for some unique cone because if $\mathcal{O}$ is any orbit then

$$\mathcal{O} = \mathcal{O}_\sigma$$

where σ is the minimal cone for which $\mathcal{O} \subset X_\sigma$. This sets up an inclusion-reversing correspondence between orbits and cones because $\dim \mathcal{O}_\sigma = n - \dim \sigma$ and $\mathcal{O}_\sigma \subset \overline{\mathcal{O}_\tau}$ if and only if $\tau < \sigma$. Furthermore, the orbits form a Whitney stratification of X_Φ.

We refer the reader to Fulton [61, §2.1,3.1] for proofs of these facts (which can readily be verified in the examples we have seen). Note that it follows from this description that the affine toric variety $X_{\langle \sigma \rangle}$ is a normal slice to the orbit $\mathcal{O}_\sigma$. The closure of $\mathcal{O}_\sigma$ can also be explicitly described as a toric variety associated to the fan Star (σ) whose cones correspond to $\{\tau \in \Phi | \tau \geq \sigma\}$.

9.6   Subdivisions and desingularisations

When is a toric variety X_Φ non-singular? This is a local question because a variety is non-singular if it is non-singular at each point. Thus we can reduce to the question: when is an affine toric variety X_σ corresponding to a rational cone σ in V non-singular? There is a simple answer:

Proposition 9.6.1 (Fulton, [61, §2.1]). *X_σ is non-singular if and only if σ is generated by a subset of some basis for the lattice L (in which case we say σ is a **non-singular cone**).*

Examples 9.6.2.   1. Let $\{e_i\}$ be the standard basis for $\mathbb{Z}^n$ and suppose σ is a cone generated by a subset of k of these basis vectors. Then σ is a product of k cones of the form $\mathbb{R}_{\geq 0} \subset \mathbb{R}$ and the 0-dimensional cone in $\mathbb{R}^{n-k}$. It follows that the affine toric variety X_σ is isomorphic to

$$\mathbb{C}^k \times (\mathbb{C}^*)^{n-k}.$$

Clearly any non-singular affine toric variety is of this kind.

2. Now consider the cone τ in $\mathbb{Z}^3$ generated by the four vectors $(1, \pm 1, \pm 1)$. The standard basis determines an identification of $\mathbb{Z}^3$ with its dual under which the cone τ is identified with its dual. It follows that M_τ is a monoid generated by four elements $f_1, \ldots, f_4$ with the single relation

$$f_1 + f_2 = f_3 + f_4.$$

The corresponding affine toric variety X_τ is therefore isomorphic to the complex cone

$$\{(X, Y, Z, W) \in \mathbb{C}^4 \mid XY = ZW\}$$

over the quadric $XY = ZW$ in $\mathbb{CP}^3$. It is easy to see that there is a singularity at the origin of the cone.

Non-singular toric varieties thus correspond to fans all of whose cones are non-singular. This is a rather strong condition and we will actually be more interested in a slightly weaker one.

Definition 9.6.3. A cone σ is **simplicial** if it is generated by linearly independent vectors. A fan is simplicial if all its cones are simplicial.

If σ is an n-dimensional simplicial rational fan with respect to L then it is generated by n elements $v_1, \ldots, v_n$. The minimal non-zero elements in $L \cap \mathbb{R}_{\geq 0} \cdot v_i$ for $i = 1, \ldots, n$ generate a sublattice L' of finite index in L. By definition the cone σ' generated by these minimal elements is a non-singular rational cone with respect to L'. The inclusion of L' in L induces a map from the non-singular variety $X_{\sigma'}$ to X_σ. Fulton [61, §2.2] shows that this is the quotient of $X_{\sigma'}$ by an action of the finite group L/L'. There is a similar picture for simplicial fans of lower dimension and we deduce

Proposition 9.6.4. *If Φ is simplicial then the toric variety X_Φ is an **orbifold**. (Very roughly an n-dimensional orbifold is a space locally modelled on a quotient of $\mathbb{R}^n$ by the action of a finite group — see Ruan [149] for details.)*

Later we will want to drop the rationality condition on the fan. If a fan is not rational it cannot be non-singular (since it is not generated by elements of the lattice at all) but it can be simplicial, and the latter turns out to be the appropriate 'non-singularity' condition for non-rational fans.

We can measure how far a fan is from being simplicial. We say an edge, that is a one-dimensional face, of a cone is **free** if all the other edges are contained in a single codimension one face. A cone is called **deficient** if it has no free edges. In these terms, a fan is simplicial if and only if it has no deficient cones.

We can always reduce the number of deficient cones by subdividing — a **subdivision** Ψ of a fan Φ is a fan with $|\Psi| = |\Phi|$ and such that each cone of Ψ is contained in a unique cone of Φ. For example, if ρ is a one-dimensional ray in the interior of a cone σ then we can define a new fan, the **star subdivision** of Φ at ρ, by

$$\mathrm{Star}\,(\Phi, \rho) = \Phi - \mathrm{Star}\,(\sigma) \cup \{\rho + \tau \mid \tau \in \partial\mathrm{Star}\,(\sigma)\}$$

where $\rho + \tau$ is the cone spanned by ρ and τ. The edge generated by ρ is free in any cone of which it is a face. It follows that if ρ is a ray in the interior of a deficient cone of maximal dimension amongst such cones then the star subdivision $\mathrm{Star}\,(\Phi, \rho)$ contains fewer deficient cones than Φ. Continuing in this way we can eliminate all the deficient cones and so obtain a simplicial subdivision.

If the fan Φ is rational then by subdividing at rational rays we obtain a rational simplicial subdivision Ψ. Since each cone of Ψ is contained in a cone of Φ the identity induces a map

$$X_\Psi \to X_\Phi.$$

By Proposition 9.6.4 X_Ψ is an orbifold and it turns out that this map is surjective and an isomorphism on a dense open subset. In fact we can further subdivide Ψ so that it is a non-singular fan — see Fulton [61, §2.6]. It follows that we can always desingularise a toric variety by subdividing the corresponding fan.

9.7   Equivariant intersection cohomology

When a group G acts on a topological space X we can define **equivariant homology** groups $H_*^G(X)$. The example we have in our sights is the torus action on a toric variety. Thus, although the theory applies much more widely, we will simplify our treatment by assuming that $G = T$ is an algebraic torus acting on a complex variety X. We will take the coefficients of our homology groups to be in the reals $\mathbb{R}$.

There are several equivalent approaches to defining equivariant homology. Perhaps the most common is to define

$$H_*^T(X) = H_*(ET \times_T X)$$

where ET is a contractible space upon which T acts freely on the right, and the Borel construction $ET \times_T X$ is the quotient of $ET \times X$ by the action

$$t \cdot (e, x) = (e \cdot t^{-1}, t \cdot x).$$

(The notation for the equivariant homology groups should not be confused with that for the homology groups with respect to a triangulation introduced in §2.1 which was, unfortunately, the same!) This is well-defined because such a space ET is unique up to T-equivariant homotopy. For a non-trivial torus the space ET is infinite-dimensional; when $T = \mathbb{C}^*$ the standard model is $\mathbb{C}^\infty - \{0\}$ where $\mathbb{C}^*$ acts via

$$\lambda \cdot (x_0, x_1, x_2, \ldots) = (\lambda x_0, \lambda x_1, \lambda x_2, \ldots).$$

More generally, for a k-dimensional torus ET is $(\mathbb{C}^\infty - \{0\})^k$. However, if we wish to avoid infinite-dimensional spaces then we can express ET as a limit of spaces $ET_n = (\mathbb{C}^n - \{0\})^k$ and put

$$H_*^T(X) = \lim H_*(ET_n \times_T X).$$

Alternatively, we can take a more geometric approach. Suppose that X has a piecewise-linear structure and that T acts via piecewise-linear maps. (This is the case, for instance, for the torus action on a projective toric variety.) A T-**equivariant simplicial** i-**chain** is determined by an $(i + \dim T)$-dimensional simplicial chain in some $\mathbb{R}^n$ (with respect to some given triangulation of R^n), together with a free action of T on the support $|\xi|$ and a T-equivaraint piecewise-linear map $f : |\xi| \to X$. Two such pairs (ξ, f) determine the same equivariant simpicial chain if they are the same up to the equivalence relation generated by subdivisions of triangulations and the standard embeddings $\mathbb{R}^n \subset \mathbb{R}^{n+1} \subset \cdots$. The boundary $(\partial\xi, f|_{|\partial\xi|})$ of an equivariant simplicial chain is again an equivaraint simplicial chain so that we obtain a complex $C_*^T(X)$ of equivariant simplicial chains. The (simplicial) T-equivariant homology of X is the homology of this complex. A version of this construction using subanalytic rather than simplicial chains is discussed in more detail in Goresky, Kottwitz and MacPherson [67].

Finally, one can define equivariant homology via sheaf theory as the cohomology groups of an object $\mathcal{S}_{T,X}^\bullet$ of the equivariant derived category of sheaves on X. The theory, developed in Bernstein and Lunts [16], is rather subtle and we will do no more here than remark its existence.

All of these definitions generalise to intersection homology. If T acts preserving a stratification of X then we can define

$$I^p H_*^T(X) = I^p H_*(ET \times_T X)$$

where $ET \times_T X$ is stratified in the obvious way with strata $ET \times_T S$ for the strata S of X, see Kirwan [112]. If we prefer a geometric to a homotopy-theoretic approach, we can say that an equivariant simplicial i-chain is p-allowable if

$$\dim f^{-1}(X_{m-k} - X_{m-k-1}) \leq i + \dim T - k + p(k),$$

and define the equivariant intersection homology as the homology of the complex $I^p S_*^T(X)$ of p-allowable equivariant chains with p-allowable boundaries — the subanalytic version of this construction is explained in Goresky, Kottwitz and MacPherson [67]. Finally, if we want to think in terms of sheaf theory then following Bernstein and Lunts [16] we can construct a complex $\mathcal{I}^p \mathcal{S}_{T,X}^\bullet$ in the equivariant derived category whose cohomology groups are the equivariant intersection homology of X.

We can define equivariant cohomology and intersection cohomology groups by dualising and henceforward we will work with these. Both the equivariant cohomology and intersection cohomology of a space are graded modules over the equivariant cohomology H_T^* of a point. This additional structure is very useful and so we digress briefly on the subject of graded rings and modules.

Definition 9.7.1. A **graded ring** is a ring R together with a decomposition

$$R = R_0 \oplus R_1 \oplus \ldots$$

into additive subgroups such that if $r \in R_i$ and $s \in R_j$ then the product $rs \in R_{i+j}$. Note, in particular, that R_0 is a subring. We say the elements of R_i have homogeneous degree i and write R^+ for the ideal generated by the homogeneous elements of strictly positive degree. Similarly, a **graded module** M over R is a module over R together with a decomposition

$$M = M_0 \oplus M_1 \oplus \ldots$$

into additive subgroups such that if $r \in R_i$ and $m \in M_j$ then the product $rm \in M_{i+j}$.

There are two constructions involving graded modules which we will require.

Definition 9.7.2. For a graded R-module M we let

- $M(k)$ be the graded module whose homogeneous degree i elements are $M(k)_i = M_{k+i}$ and,

- $\overline{M}$ be the quotient M/R^+M. This is naturally a module over $R/R^+ \cong R_0$ and inherits a grading from M (which is preserved by the action of R_0).

The example we have in mind is the graded ring H_T^* where T is a k-dimensional torus. This is the cohomology ring

$$H^*\left((\mathbb{C}^\infty)^k \times_T \{\text{pt}\}\right) \cong H^*\left((\mathbb{CP}^\infty)^k\right) \cong (H^*(\mathbb{CP}^\infty))^{\otimes k}.$$

The cohomology of $\mathbb{CP}^\infty$ is the polynomial ring over $\mathbb{R}$ with one generator of degree 2 and so H_T^* can be identified with the polynomial ring in n variables of degree 2. The structure of the equivariant intersection cohomology $IH_T^*(X)$ as a module over this ring in many important cases is very simple. We say X is **equivariantly formal** if there is a (non-canonical) isomorphism

$$IH_T^*(X) \cong H_T^* \otimes_\mathbb{R} IH^*(X),$$

i.e. if $IH_T^*(X)$ is a free module over the graded ring H_T^* generated by a basis for $IH^*(X)$. (This occurs if and only if the Leray spectral sequence for the fibration

$$X \longrightarrow ET \times_T X$$
$$\downarrow$$
$$ET/T$$

degenerates.)

Proposition 9.7.3 (Goresky, Kottwitz and MacPherson [67, Thm. 14.1]). *If X is a complex projective variety, or if all the odd-dimensional intersection homology of X vanishes, then X is equivariantly formal.*

For an equivariantly formal space we note that we can recover the intersection cohomology from the equivariant intersection cohomology as

$$IH^*(X) \cong \overline{IH_T^*(X)}$$

where $\overline{IH_T^*(X)}$ is constructed from the graded H_T^*-module $IH_T^*(X)$ as in Definition 9.7.2.

For (non-equivariant) intersection cohomology we have a local calculation relating the intersection cohomology of a cone to the intersection cohomology of the cone less the vertex. This allows us to obtain results inductively by adding a stratum at a time. There is no direct analogue for equivariant intersection cohomology and in general the situation is more complicated. However, Bernstein and Lunts [16] prove an important result which plays a similar rôle in the cases in which we are interested. Before we state it we need to introduce some terminology.

Definition 9.7.4. If M is a module over a ring R then a **projective cover** of M is a projective module P together with a surjection onto M such that any other surjection $P' \to M$ from a projective module factorises uniquely through a surjection $P' \to P$, i.e.

$$P \longrightarrow\!\!\!\!\!\longrightarrow M.$$

Projective covers exist and are unique up to isomorphism.

The projective cover of a finitely-generated H_T^*-module M is easy to construct. Since H_T^* is a principal ideal domain, projective and free modules will be the same thing, so we can replace 'projective' by 'free' in the above definition. We let $P = H_T^* \otimes_{\mathbb{R}} \overline{M}$, where $\overline{M}$ is constructed as in Definition 9.7.2. Then we choose a linear, grading-preserving, section $s : \overline{M} \to M$ of the quotient $M \to \overline{M}$ and define a map $\pi : P \to M$ by

$$\pi(a \otimes m) = a \cdot s(m).$$

We need to check that this is a surjection: suppose $v \in M$ and write $\overline{v}$ for the residue class in $\overline{M}$. Recalling that s is a section we have

$$\overline{v - s(\overline{v})} = 0.$$

It follows that we can find $a \in H_T^*$ and $u \in M$ such that

- $\deg u < \deg v$;

- $v - s(\overline{v}) = a \cdot u$.

Since M is finitely-generated as a module over H_T^* we may assume inductively that any element of M with degree strictly less than that of v is in the image of the map $\pi : P \to M$. In particular we have $u = \pi(b \otimes \overline{w})$ for some $b \in H_T^*$ and $w \in M$. It follows that v is in the image of π too, because

$$v = s(\overline{v}) + a \cdot u = \pi(1 \otimes \overline{v} + ab \otimes \overline{w}).$$

We leave the reader to verify that P is the projective cover of M.

Theorem 9.7.5 (Bernstein and Lunts, [16, §14]). *Suppose T acts algebraically on an affine variety V with a fixed point v. Furthermore, suppose that there is a 1-dimensional subtorus $\mathbb{C}^* \hookrightarrow T$ such that v is the unique fixed point of the $\mathbb{C}^*$-action and that v is an attraction point, i.e. that $\lambda \cdot w \to v$ as $\lambda \to 0$ for any $w \in V$. Then $IH_T^*(V)$ is the projective cover of $IH_T^*(V - \{v\})$ via the restriction map.*

9.8   The intersection cohomology of fans

In this section we will show how the equivariant intersection cohomology of a toric variety may be computed from combinatorial data. When the toric variety is affine or compact we can also compute the intersection cohomology. These ideas can be generalised to define a notion of intersection cohomology for fans.

Suppose that Φ is a fan in a real vector space V of dimension n.

Definition 9.8.1. We make Φ into a topological space, the **fan space** which we also denote by Φ, by decreeing that a subset $U \subset \Phi$ is open if and only if $\sigma \in U$ and $\tau < \sigma$ implies $\tau \in U$. More succinctly: the open subsets are the subfans.

The closure of the point σ is Star $(\sigma) = \{\tau \geq \sigma\}$. The point σ is contained in a smallest open set; namely the open set $[\sigma]$ corresponding to the subfan generated by σ. It follows that the stalks of a sheaf $\mathcal{E}$ on the fan space Φ are

$$\mathcal{E}_\sigma = \operatorname{colim}_{U \ni \sigma} \mathcal{E}(U) = \mathcal{E}([\sigma]).$$

Furthermore, the open sets $[\sigma]$ form a basis for the topology and so $\mathcal{E}$ is determined completely by the assignment $\sigma \mapsto \mathcal{E}_\sigma$ and the restriction maps $\mathcal{E}_\sigma \to \mathcal{E}_\tau$ for $\tau < \sigma$.

Now suppose that Φ is rational. Then the fan space is (homeomorphic to) the quotient of the associated toric variety X_Φ by the natural action of a torus T. This follows immediately from the description of the orbits of a toric variety X_Φ in Proposition 9.5.1 with one orbit $\mathcal{O}_\sigma$ for each cone σ in Φ. Let $q : X_\Phi \to X_\Phi/T \cong \Phi$ be the quotient map.

There is an equivariant complex of sheaves $\mathcal{IS}^\bullet_{T,X_\Phi}$ on X_Φ whose cohomology groups are the equivariant intersection cohomology groups $IH^*_T(X_\Phi)$. Consider the push-forward

$$Rq_* \mathcal{IS}^\bullet_{T,X_\Phi}$$

of this complex to Φ. We define a sheaf of graded vector spaces $\mathcal{L}_\Phi$ whose i^{th} graded piece is the cohomology sheaf $\mathcal{H}^i(Rq_* \mathcal{IS}^\bullet_{T,X_\Phi})$. Since $[\sigma]$ is the smallest open set containing σ the stalk of $\mathcal{L}_\Phi$ at a cone σ is

$$\mathcal{L}_{\Phi,\sigma} = IH^*_T(q^{-1}[\sigma]) = IH^*_T(X_\sigma) = IH^*_T(T \times_{T_\sigma} X_{\langle\sigma\rangle}) = IH^*_{T_\sigma}(X_{\langle\sigma\rangle})$$

where we have used the identification $X_\sigma \cong T \times_{T_\sigma} X_{\langle\sigma\rangle}$ from §9.5. The sheaf $\mathcal{L}_\Phi$ has additional structure. The stalk $\mathcal{L}_\sigma$ is a module over $H^*_{T_\sigma}$ and these module structures on the stalks are compatible. What this amounts to is that $\mathcal{L}_\Phi$ is a module over a certain sheaf of graded rings $\mathcal{A}_\Phi$ with stalks

$$\mathcal{A}_{\Phi,\sigma} = H^*_{T_\sigma}. \tag{9.2}$$

(By a module over $\mathcal{A}_\Phi$, or an $\mathcal{A}_\Phi$-module, we mean a sheaf $\mathcal{M}$ on Φ whose sections over an open set U form a graded module over the ring $\mathcal{A}_\Phi(U)$ compatibly with the restriction maps. By this we mean that for any open sets $U \supset V$ there is a commutative diagram

$$
\begin{array}{ccc}
\mathcal{A}_\Phi(U) \times \mathcal{M}(U) & \longrightarrow & \mathcal{A}_\Phi(V) \times \mathcal{M}(V) \\
\downarrow & & \downarrow \\
\mathcal{M}(U) & \longrightarrow & \mathcal{M}(V),
\end{array}
$$

in which the horizontal arrows are restrictions and the vertical ones the action of $\mathcal{A}_\Phi$.)

Using the description of the equivariant cohomology of a point as the ring of polynomial functions on a vector space we can describe $\mathcal{A}_\Phi$ as follows: the sections over an open set U are the cone-wise polynomial functions on the subset

$$|U| = \cup_{\sigma \in U} |\sigma|$$

of V. The restriction maps of $\mathcal{A}_\Phi$ arise from restriction of functions. The stalk $\mathcal{A}_{\Phi,\sigma}$ at a point σ is the ring of polynomial functions on $|\sigma|$ or, equivalently, the ring of polynomial functions on $\mathrm{Span}(\sigma)$. If we grade these so that the linear functions have degree *two* then we obtain (9.2).

Theorem 9.8.2 (Theorem A). *1. $\mathcal{L}_\Phi$ is a* **flabby** *sheaf, i.e. the restriction map $\mathcal{L}_\Phi(\Phi) \to \mathcal{L}_\Phi(U)$ from the global sections is a surjection for every open subset U.*

2. The oddly graded pieces of the stalks $\mathcal{L}_{\Phi,\sigma}$ all vanish.

Theorem 9.8.3 (Theorem B). *For any subfan Ψ of the rational fan Φ there is an isomorphism of graded H_T^*-modules*

$$IH_T^*(X_\Psi) \cong \mathcal{L}_\Phi(\Psi).$$

(Here the global sections $\mathcal{L}_\Phi(\Psi)$, which naturally form an $A_\Phi(\Psi)$-module, are made into an H_T^*-module via the ring map $H_T^* \to A_\Phi(\Psi)$ arising from restriction of a polynomial function on V to a polynomial function on $|\Psi|$.)

We prove these theorems inductively in tandem using the following scheme:

Theorem A for $\Phi_{<k}$ $\Longrightarrow$ Theorem B for $\Phi_{<k}$ $\Longrightarrow$ Theorem A for $\Phi_{\leq k}$

where $\Phi_{<k}$ is the subfan

$$\{\sigma \in \Phi \mid \dim \sigma < k\}.$$

consisting of cones of dimension $< k$, and $\Phi_{\leq k}$ is defined similarly.

Assume Theorem A holds for $\Phi_{<k}$. Then each stalk of $\mathcal{L}_{\Phi_{<k}}$ is a graded vector space whose oddly-graded pieces vanish. We can consider $\mathcal{L}_{\Phi_{<k}}$ as a complex of sheaves concentrated in even degrees with zero differentials. The terms are the cohomology sheaves of the restriction of $Rq_*\mathcal{IS}_{T,X_\Phi}^\bullet$ to $\Phi_{<k}$. Straightforward homological algebra shows that the vanishing of the odd degree terms means that $\mathcal{L}_{\Phi_{<k}}$ is quasi-isomorphic to the restriction of $Rq_*\mathcal{IS}_{T,X_\Phi}^\bullet$ to $\Phi_{<k}$.

It is a standard result that the higher cohomology of a flabby sheaf vanishes, see for example Iversen [84, II.3.5]. Thus the ith hypercohomology of $\mathcal{L}_{\Phi_{<k}}$ (thought of as a complex of flabby sheaves) over a subfan Ψ is simply the sections over Ψ of the ith term in the complex. Combining this with the above quasi-isomorphism we have an isomorphism of graded vector spaces

$$\mathcal{L}_\Phi(\Psi) \cong H^*(\Psi; \mathcal{L}_\Phi) \cong H^*(\Psi; Rq_*\mathcal{IS}_{T,X_\Phi}^\bullet) \cong H^*(X_\Psi; \mathcal{IS}_{T,X_\Phi}^\bullet) \cong IH_T^*(X_\Psi)$$

for any subfan Ψ of $\Phi_{<k}$. We can check that this is actually an isomorphism of graded H_T^*-modules and hence we have proved Theorem B for $\Phi_{<k}$.

In order to prove Theorem A for $\Phi_{\leq k}$ we study the restriction map

$$\mathcal{L}_{\Phi,\sigma} = \mathcal{L}_\Phi([\sigma]) \to \mathcal{L}_\Phi(\partial\sigma) \tag{9.3}$$

for a cone σ of dimension k. We know that the stalk $\mathcal{L}_{\Phi,\sigma}$ is the equivariant intersection cohomology group

$$IH_T^*(X_\sigma)$$

and, by Theorem B, that the sections $\mathcal{L}_\Phi(\partial\sigma)$ are isomorphic to

$$IH_T^*(X_{\partial\sigma}).$$

Assume for a moment that $k = \dim V$. Then X_σ has a unique fixed point x_σ and $X_{\partial\sigma}$ is the complement $X_\sigma - \{x_\sigma\}$. Thus the restriction map can be identified with the restriction

$$IH_T^*(X_\sigma) \to IH_T^*(X_\sigma - \{x_\sigma\}).$$

If $k < \dim V$ then we can use the identification $X_\sigma = T \times_{T_\sigma} X_{\langle\sigma\rangle}$ from §9.5 to obtain a similar result where we replace T by T_σ and X_σ by $X_{\langle\sigma\rangle}$.

The hypotheses of Theorem 9.7.5 are satisfied and we deduce that the $\mathcal{A}_{\Phi,\sigma}$-module $\mathcal{L}_{\Phi,\sigma}$ is the projective cover of $\mathcal{L}_\Phi(\partial\sigma)$. In particular, the restriction (9.3) is surjective. The latter is equivalent to flabbiness for we have

Lemma 9.8.4. *A sheaf $\mathcal{E}$ on Φ is flabby if and only if the restriction maps*

$$\mathcal{E}_\sigma \to \mathcal{E}(\partial\sigma) \tag{9.4}$$

are surjective for all cones σ.

Proof. If $\mathcal{E}$ is flabby then (9.4) follows immediately from the commutative diagram

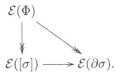

Conversely, suppose (9.4) holds for all σ. Then (9.4) guarantees that we can extend a section over the open set $U \cup \Phi_{\leq k}$ to one over $U \cup \Phi_{\leq k+1}$. By induction we can then extend a section over $U = U \cup \Phi_{\leq 0}$ to one over $\Phi = U \cup \Phi_{\leq n}$. $\square$

Finally, note that the oddly-graded pieces of the stalks of $\mathcal{L}_\Phi$ at points $\tau \in \partial\sigma$ vanish (by Theorem A for $\Phi_{<k}$) which implies that the oddly-graded pieces of

$$\mathcal{L}_\Phi(\partial\sigma)$$

vanish. It is easy to see that the same then holds for the projective cover $\mathcal{L}_{\Phi,\sigma}$. This completes the proofs of Theorems A and B.

Combining Theorems A, B and Theorem 9.7.3 we obtain the following corollary which enables us to compute the intersection cohomology of affine and projective toric varieties.

Corollary 9.8.5. *The affine toric variety X_σ associated to a rational cone is equivariantly formal, as is the projective toric variety associated to a projective rational fan Φ. Hence the equivariant intersection cohomology of such a variety is a free H_T^*-module generated by a basis for the intersection cohomology. In particular, for any cone in a rational fan Φ we have*

$$IH^*(X_\sigma) \cong \overline{\mathcal{L}_{\Phi,\sigma}}$$

and if the fan is projective (i.e. complete and equipped with an ample function) we have $IH^*(X_\Phi) \cong \overline{\mathcal{L}_\Phi(\Phi)}$.

If we abstract the properties of $\mathcal{L}_\Phi$ then we are led to the following definition, where now Φ is an arbitrary, not necessarily rational, fan.

Definition 9.8.6. A sheaf $\mathcal{M}$ on Φ is a **minimal extension sheaf** if $\mathcal{M}$ is an $\mathcal{A}_\Phi$-module with stalk $\mathcal{M}_{\{0\}} = \mathbb{R}$ at the origin and such that

$$\mathcal{M}_\sigma \to \mathcal{M}(\partial\sigma)$$

is the projective cover for each cone σ.

Theorem 9.8.7. *For any fan Φ there is a unique minimal extension sheaf (up to isomorphism). For a rational fan it agrees with $\mathcal{L}_\Phi$ as defined above.*

Proof. A minimal extension sheaf can be constructed inductively; if we have already defined $\mathcal{M}$ over the subfan $\Phi_{<k}$ then we let the stalk $\mathcal{M}_\sigma$ at a cone of dimension k be the projective cover of $\mathcal{M}(\partial\sigma)$ and use the surjection

$$\mathcal{M}_\sigma \to \mathcal{M}(\partial\sigma)$$

to define the restriction maps. The uniqueness follows from the uniqueness of projective covers. We have already seen that the sheaf $\mathcal{L}_\Phi$ on a rational fan satisfies the conditions. $\square$

Example 9.8.8. If Φ is a simplicial fan then the graded sheaf $\mathcal{A}_\Phi$ is the minimal extension $\mathcal{L}_\Phi$ constructed above. To see that $\mathcal{A}_\Phi$ is a minimal extension sheaf for simplicial Φ we note that for any simplicial cone σ the restriction

$$\mathcal{A}_{\Phi,\sigma} \to \mathcal{A}_\Phi(\partial\sigma)$$

is surjective. This is equivalent to the statement that, given a set of k linearly independent vectors in $\mathbb{R}^k$ and real polynomials defined on the hyperplanes spanned by the subsets of size $k - 1$, whose restrictions to the intersections agree, there is a polynomial defined on the entirety of $\mathbb{R}^k$ which extends the given ones. The corresponding statement is false if the vectors are linearly dependent; there are some sets of polynomials which do not have a common extension to $\mathbb{R}^k$. It follows that if Φ is not simplicial then $\mathcal{A}_\Phi$ is not a minimal extension sheaf.

Minimal extension sheaves have good properties. They are flabby and are **locally-free** as $\mathcal{A}_\Phi$-modules, i.e. the stalks are finitely-generated free modules over the graded ring $\mathcal{A}_{\Phi,\sigma}$. These properties follow directly from the properties of projective covers. The terminology arises because the minimal extension sheaf is (in a precise sense) the 'smallest' flabby, locally-free sheaf with stalk $\mathbb{R}$ at the origin.

We can extend the definition to say that a **minimal extension sheaf based at** σ is a sheaf $\mathcal{M}$ on Φ which is supported on the closure $\mathrm{Star}(\sigma)$ of a cone, has stalk $\mathcal{M}_\sigma = \mathbb{R}$ and is such that

$$\mathcal{M}_\tau \to \mathcal{M}_{\partial\tau}$$

is the projective cover for each $\tau > \sigma$. We denote such a sheaf by $\mathcal{L}_{\mathrm{Star}(\sigma)}$. This agrees with our earlier notation since $\mathrm{Star}(\{0\}) = \Phi$.

By analogy with the situation for rational fans we make the following definition.

Definition 9.8.9 (Bressler and Lunts [32, Defn. 5.7]). The **intersection cohomology** $IH(\Phi)$ **of a fan** Φ is the graded vector space

$$\overline{\mathcal{L}_\Phi(\Phi)} = \mathcal{L}_\Phi(\Phi)/A^+ \cdot \mathcal{L}_\Phi(\Phi)$$

where A is the graded ring of polynomial functions on $|\Phi|$.

It follows from Corollary 9.8.5 that if Φ is a projective fan we have

$$IH(\Phi) = IH^*(X_\Phi).$$

Many of the theorems which hold for the intersection cohomology of a toric variety can be generalised to the context of the intersection cohomology of a fan. In particular there are analogues of the decomposition theorem and the hard Lefschetz theorem.

Theorem 9.8.10 (Decomposition theorem for fans, Bressler and Lunts [32, Thm. 5.3 and 5.5]). *If* $\pi : \Psi \to \Phi$ *is a subdivision of a complete fan then the push-forward* $R\pi_* \mathcal{L}_\Psi$ *is quasi-isomorphic to a direct sum each of whose terms is a minimal extension sheaf* $\mathcal{L}_{\mathrm{Star}(\sigma)}(m_\sigma)$ *based at a cone* σ *with grading shifted by an integer* m_σ. *One of the terms will be the minimal extension sheaf* $\mathcal{L}_\Phi$. *In particular, there is a direct sum decomposition*

$$IH(\Psi) \cong IH(\Phi) \oplus \bigoplus_{i \in I} IH\left(\mathrm{Star}(\sigma_i)\right)(m_{\sigma_i}). \qquad (9.5)$$

It is important to note that this theorem is *much* easier to prove than the decomposition theorem for a proper map of complex varieties.

Theorem 9.8.11 (Hard Lefschetz theorem for fans). *Suppose* Φ *is a complete fan in a vector space* V *and* $\ell : |\Phi| \to \mathbb{R}$ *is an ample function, i.e. a strictly convex, cone-wise linear continuous function. Then* ℓ *is a degree two element in* $\mathcal{A}_\Phi(\Phi)$ *and so multiplication by* ℓ *induces a degree two map of the graded vector space* $IH(\Phi)$. *The composite*

$$\ell^i : IH^{n-i}(\Phi) \to IH^{n+i}(\Phi)$$

is an isomorphism for $0 \leq i < n = \dim V$.

This was conjectured by Bressler and Lunts [32] and then proved by Karu [90]. Subsequently the proof was streamlined by Bressler and Lunts [33]. The proof involves reducing to the case of a simplicial fan for which the result was proved by McMullen [129]. This reduction is achieved by noting that every fan has a simplicial subdivision and then carefully analysing the inclusion of $IH(\Phi)$ into the intersection cohomology of a subdivision arising from the decomposition (9.5).

An important consequence of the hard Lefschetz theorem for fans is that we can use it to obtain a combinatorial formula for the intersection Betti numbers of a complete fan (Bressler and Lunts [32, §7]). It tells us that the ith Betti number

$$\dim IH^i(\Phi)$$

is the coefficient of t^i in a polynomial $h(\Phi, t)$ which is defined recursively by

$$h(\Phi, t) = \sum_{\sigma \in \Phi} g(\sigma, t)(t^2 - 1)^{\dim V - \dim \sigma} \qquad (9.6)$$

where $g(\sigma, t)$ is the polynomial with coefficients $g_i(\sigma)$ given by

$$g_i(\sigma) = \begin{cases} h_i(\overline{\partial \sigma}) - h_{i-2}(\overline{\partial \sigma}) & 0 \leq i < \dim \sigma \\ 0 & \text{otherwise.} \end{cases}$$

Here the fan $\overline{\partial \sigma}$ is the projection of the fan $\partial \sigma$ onto a subspace $W \subset \mathrm{Span}(\sigma)$ where $\mathrm{Span}(\sigma) = W \oplus \langle \rho \rangle$ for some ρ in the interior of σ. (Effectively we take the boundary of σ and flatten it out.) $\overline{\partial \sigma}$ is always a complete fan and the graph of $\partial \sigma$ defines an ample function on it. To get the recursion off the ground we put $h(\{0\}, t) = 1 = g(\{0\}, t)$.

In geometric terms (which can be made precise for rational fans) the intersection cohomology of the fan Φ arises from local contributions from each of the 'affine open subsets' corresponding to cones. The contribution from a cone σ is the 'primitive part' of the intersection cohomology of the 'projective variety' associated to the complete cone $\overline{\partial \sigma}$ (cf. the discussion of the hard Lefschetz theorem in §4.10).

Exercise 9.8.12. Prove that if σ is a cone over a k-dimensional simplex then

$$h(\overline{\partial \sigma}, t) = \frac{t^{2(k+1)} - 1}{t^2 - 1}.$$

Deduce that $g(\sigma, t) = 1$.

Remark 9.8.13. The intersection Betti numbers of the toric variety X_Φ corresponding to a complete rational fan Φ can be computed recursively in this way. Note that they only depend on the combinatorics of the cones and not on how they sit as subsets of V (information which certainly enters into the construction of X_Φ). This is not true of the ordinary Betti numbers if the polytope is not simplicial (Stanley [169, p20]).

Example 9.8.14. As an example we compute the intersection Betti numbers of the toric variety X_Q where Q is the octahedron with vertices at $(\pm 1, 0, 0)$, $(0, \pm 1, 0)$ and $(0, 0, \pm 1)$. This is the toric variety associated to the fan Φ_Q over the polar polytope, which is the cube with vertices $(\pm 1, \pm 1, \pm 1)$. The

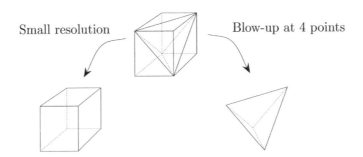

Small resolution   Blow-up at 4 points

Figure 9.5: The fan over a cube can be desingularised by subdividing at an inscribed tetrahedron. The corresponding map of toric varieties is a small resolution. The cube with inscribed tetrahedron is also a subdivision of the tetrahedron corresponding to the blow-up of $\mathbb{CP}^3$ at four points.

variety is obtained by glueing six copies (one for each face of the cube) of the complex cone over the quadric $XY = ZW$ which we met in Examples 9.6.2. We deduce that X_Q is a 3-fold with six singular points.

We need to compute the polynomial $g(\sigma, t)$ for each cone $\sigma \in \Phi_Q$. We saw above that $g(\sigma, t) = 1$ when σ is a cone over a simplex. It remains to compute $g(\sigma, t)$ for the cone over a face of the cube. It is easy to see that $\overline{\partial \sigma}$ is a complete fan in $\mathbb{R}^2$ with four 2-dimensional cones so that we have

$$h(\overline{\partial \sigma}, t) = (t^2 - 1)^2 + 4(t^2 - 1) + 4 = t^4 + 2t^2 + 1$$

and $g(\sigma, t) = 1 + t^2$. It follows that

$$
\begin{aligned}
h(\Phi_Q, t) &= (t^2 - 1)^3 + 8(t^2 - 1)^2 + 12(t^2 - 1) + 6(1 + t^2) \\
&= 1 + 5t^2 + 5t^4 + t^6.
\end{aligned}
$$

We can verify this result by giving a second description of X_Q. Inscribe a tetrahedron in the cube as shown in Figure 9.5. The fan Ψ over the resulting polytope is a subdivision of Φ_Q representing a desingularisation. The fibre of the map

$$X_\Psi \to X_Q$$

over each singular point is a $\mathbb{CP}^1$ and so this is a small resolution (in the sense of Definition 8.4.6). In particular

$$H^*(X_\Psi; \mathbb{R}) \cong IH^*(X_Q; \mathbb{R}).$$

The fan Ψ is also a subdivision of the fan over the tetrahedron, which corresponds to the toric variety $\mathbb{CP}^3$. The map

$$X_\Psi \to \mathbb{CP}^3$$

is the blow-up of $\mathbb{CP}^3$ at the four fixed points of the torus action. Hence we have

$$h(\Phi_Q, t) = h(\Psi, t) = h(\mathbb{CP}^3, t) + 4\left(h(\mathbb{CP}^2, t) - 1\right) = 1 + 5t^2 + 5t^4 + t^6$$

as before.

9.9   Stanley's conjectures

Suppose P is an n-dimensional convex polytope. Define the **face vector** or f-vector to be $(f_0(P), \ldots, f_{n-1}(P))$ where $f_i(P)$ is the number of i-dimensional faces of P. By convention we set $f_{-1}(P) = 1$ (taking the empty face to have dimension -1). A basic combinatorial question is to characterise the possible f-vectors of convex polytopes. In general, this is intractable. However, if P is a **simplicial polytope**, i.e. all the faces are simplices, then necessary and sufficient conditions, first conjectured by McMullen [128], are known for a vector to be the face vector. The conditions are given in terms of the h-**vector** $(h_0(P), \ldots, h_n(P))$ which is related to the f-vector by

$$\sum_{i=0}^{n} f_{i-1}(P)(x-1)^{n-i} = \sum_{i=0}^{n} h_i(P)x^{n-i}. \tag{9.7}$$

It should be clear that this contains the same information as the face vector.

Theorem 9.9.1 (The g-theorem). *A sequence* $(h_0, \ldots, h_n)$ *of integers is the h-vector of an n-dimensional simplicial polytope if and only if*

1. *the Dehn–Somerville relations* $h_i = h_{n-i}$ *hold for* $0 \leq i \leq \lfloor \frac{n}{2} \rfloor$,

2. *we have* $0 \leq h_0 \leq h_1 \leq \ldots \leq h_{\lfloor \frac{n}{2} \rfloor}$ *and,*

3. *the differences* $h_i - h_{i-1}$ *do not grow too fast; to be precise if*

$$n_i > n_{i-1} > \ldots > n_k \geq k \geq 1$$

   are the unique natural numbers such that

$$h_i - h_{i-1} = \binom{n_i}{i} + \binom{n_{i-1}}{i-1} + \ldots + \binom{n_k}{k}$$

   (this is not a condition — any natural number can be written uniquely in this way) then

$$h_{i+1} - h_i \leq \binom{n_i + 1}{i+1} + \binom{n_{i-1}+1}{i} + \ldots + \binom{n_k+1}{k+1}$$

   for $1 \leq i \leq \lfloor \frac{n}{2} \rfloor - 1$.

(The name of the theorem arises because the difference $h_i - h_{i-1}$ is sometimes denoted by g_i.) When $n = 2$ these conditions reduce to the single condition $f_0 = f_1$. We can also check that they hold for the tetrahedron, octahedron and icosahedron (but not, for instance, for the cube). It is also suggestive of a topological interpretation to note that in three dimensions the condition $h_0 = h_3$ is equivalent to Euler's formula

$$f_0 - f_1 + f_2 = 2.$$

Given an f-vector satisfying these conditions Billera and Lee give an explicit construction of a convex simplicial polytope with that face vector [17, 18]. The necessity of the conditions was proved by Stanley [166]. The idea is to interpret the h-vector in terms of the topology and geometry of a projective toric variety associated to P. The first step is to perturb the vertices of P so that it is rational. We can do this without changing the face vector *because P is simplicial.* Thus we may assume that P is rational, and also that the origin is in the interior. The fan over P then defines a projective toric variety. Slightly confusingly, as a result of our earlier (and standard) convention, this variety is the toric variety X_{P° associated to the *polar* polytope P°. For ease of notation we will write Y_P for X_{P°.

We can compute the intersection Betti numbers using (9.6). Recall that $g(\sigma, t) = 1$ for a cone σ over a simplex. Since i-dimensional cones of the fan over P correspond to $(i - 1)$-dimensional faces of P it follows that the intersection Betti numbers of the toric variety Y_P are the coefficients of the polynomial

$$\sum_{i=0}^{n} f_{i-1}(P)(t^2 - 1)^{\dim P - i},$$

i.e. we have $h_i = \dim IH^{2i}(Y_P; \mathbb{R})$. We know that the intersection cohomology of the projective variety Y_P satisfies the hard Lefschetz theorem. Thus there is a map

$$L : IH^*(Y_P; \mathbb{R}) \to IH^{*+2}(Y_P; \mathbb{R})$$

given by multiplication by a class $[\omega] \in H^2(Y_P; \mathbb{R})$ which induces isomorphisms

$$L^i : IH^{m-i}(Y_P; \mathbb{R}) \to IH^{m+i}(Y_P; \mathbb{R}) \tag{9.8}$$

for $0 \leq i \leq m$. In particular

$$L : IH^i(Y_P; \mathbb{R}) \to IH^{i+2}(Y_P; \mathbb{R}) \tag{9.9}$$

must be an injection for $i < m$. The first two conditions of the g-theorem are immediate consequences of (9.8) and (9.9) respectively. The third condition is more subtle. Since P is simplicial the variety Y_P is an orbifold. Hence the intersection cohomology (with real coefficients) is isomorphic to the ordinary cohomology. The third condition follows from the fact that the cohomology

$H^*(Y_P; \mathbb{R})$ is generated as a ring by classes of divisors in $H^2(Y_P; \mathbb{R})$ — see Fulton [61, §5.6] and Stanley [169, §III.1].

In this proof of the g-theorem deep results in algebraic geometry furnish us with results about the elementary (in the sense of easy to state) combinatorics of polytopes. Alternatively, a purely combinatorial proof of the g-theorem is known (McMullen [129, 130]) so that we can reverse the implications and deduce information about the intersection cohomology of toric varieties from combinatorics.

If P is not simplicial then it is no longer necessarily possible to perturb the vertices to obtain a rational polytope whilst preserving the face vector. (The non-rational polytopes form a strictly bigger class than the rational ones: there are non-rational polytopes whose combinatorial type is distinct from that of any rational polytope, see Ziegler [181, Ex. 6.21].) However there is still a complete fan Ψ_P over P. The intersection Betti numbers of this fan are the coefficients of the polynomial $h(\Psi_P, t)$ defined at (9.6) and are easily seen to be zero in odd dimensions. Stanley [167] defined the **generalised h-vector** $h(P) = (h_0(P), \ldots, h_n(P))$ of an n-dimensional polytope to consist of the even dimensional Betti numbers, i.e.

$$h_i(P) = \dim IH^{2i}(\Psi_P). \tag{9.10}$$

We saw above that this agrees with the original definition of the h-vector for simplicial polytopes but in general it is given in terms of the combinatorics of the faces by the more complicated recursive definition (9.6).

Stanley [167] proved that the generalised h-vector always satisfies 'Poincaré duality', i.e. that $h_i = h_{n-i}$ for $0 \le i \le \lfloor \dim P/2 \rfloor$. In the same paper he also conjectured that $h_i \ge 0$ for all i and that

$$h_0 \le h_1 \le \ldots \le h_{\lfloor \dim P/2 \rfloor}, \tag{9.11}$$

i.e. that the generalised h-vector has the 'hard Lefschetz property'. This conjecture was the stimulus for the work on the intersection cohomology of fans described in the previous section, and it gained the status of a theorem when Karu proved the hard Lefschetz theorem for fans in [90]; in fact Karu proved the stronger result that the Hodge–Riemann bilinear relations hold. In brief: for a complete n-dimensional fan Φ Karu constructs a Poincaré duality pairing

$$(\, , \,) : IH^{n-i}(\Phi) \times IH^{n+i}(\Phi) \to \mathbb{R}.$$

Using the Lefschetz operator $\ell^i : IH^{n-i}(\Phi) \to IH^{n+i}(\Phi)$ we obtain a bilinear form $\beta(x, y) = (\ell^i x, y)$ on $IH^{n-i}(\Phi)$. The Hodge–Riemann bilinear relations state that the form

$$(-1)^{(n-i)/2}\beta$$

is positive definite on the primitive part, i.e. on the kernel of

$$\ell^{i+1} : IH^{n-i}(\Phi) \to IH^{n+i+2}(\Phi).$$

This result was proved for simplicial fans in McMullen [129] (see also Timorin [172]). Karu's proof involves making a choice of simplicial subdivision of Φ and defining the pairing in terms of the known one on a simplicial fan. Bressler and Lunts [33] simplify the proof by showing that there is a canonical pairing (which agrees with Karu's).

9.10 Further reading

A good survey of the g-theorem, Stanley's conjectures and their proofs can be found in [169]. Stanley's book [168] puts the g-theorem in its combinatorial context, and Fulton's [61] in the geometric one of toric varieties.

General treatments of equivariant intersection cohomology can be found in Joshua [87], Brylinski [38], Bernstein and Lunts [16] and Goresky, Kottwitz and MacPherson [67]. The sheaf-theoretic viewpoint developed in Bernstein and Lunts' book culminates in a treatment of the intersection cohomology of toric varieties. The theory of sheaves on fans, and in particular of minimal extension sheaves, is developed in a similar spirit in Bressler and Lunts [32]. They give a detailed treatment of the analogue of Verdier duality, leading to a proof of Poincaré duality for the intersection cohomology of fans. As a consequence they prove the Kalai conjecture, made in [89], which states that for a cone $\tau > \sigma$ there is a coefficient-by-coefficient inequality

$$g(\sigma, t) \geq g(\tau, t)g(\mathrm{Star}(\tau), t).$$

This was initially proved for rational fans by Braden and MacPherson in [30].

An alternative treatment of sheaves on fans, combinatorial Verdier duality and Poincaré pairings which avoids the use of derived categories can be found in the series [7, 8] and [9] of papers by Barthel, Brasselet, Fieseler and Kaup.

Braden and MacPherson [31] uses similar ideas to those in §9.8 to compute equivariant intersection cohomology and intersection cohomology for a class of varieties with torus actions which includes Schubert varieties in flag varieties. The structure of the 0 and 1-dimensional orbits is used to define a 'moment graph'. There is a canonical sheaf on the moment graph, akin to the minimal extension sheaf, which can be used to compute equivariant intersection cohomology.

Chapter 10

Characteristic p and the Weil conjectures

In 1973 Deligne completed the proof of the famous Weil conjectures which relate the arithmetic of projective varieties defined over finite fields and the homology of non-singular complex projective varieties (Deligne [53, 54]). The conjectures were stated by André Weil in the 1940s (Weil [178]) and progress (leading to partial proof) was made by Grothendieck and others in the early 1960s.

10.1   Statement of the Weil conjectures

Let $X \subseteq \mathbb{CP}^n$ be a non-singular m-dimensional complex projective variety defined over an algebraic number ring R (e.g. $R = \mathbb{Z}$). That is, X can be defined by the vanishing of homogeneous polynomials with coefficients in R.

Example 10.1.1. The Fermat curve of degree n is defined in $\mathbb{CP}^2$ by the equation
$$x^n + y^n = z^n$$
over $\mathbb{Z}$.

Let π be a maximal ideal of R. (For example if $R = \mathbb{Z}$ then $\pi = p\mathbb{Z}$ where p is a prime.) Then R/π is a finite field. Let p be the characteristic of R/π. Then
$$R/\pi = \mathbb{F}_q$$
is a field with $q = p^s$ elements for some positive integer s.

We can define a projective variety
$$X \subseteq \mathbb{P}^N(\mathbb{F}_q) = \frac{\mathbb{F}_q^{N+1} - \{0\}}{\mathbb{F}_q - \{0\}}$$

163

by reducing modulo π the equations with coefficients in R which define X. If we choose π so that the characteristic p of R/π is not one of finitely many "bad" primes for X then X_π is a non-singular m-dimensional projective variety over the field $\mathbb{F}_q$.

Let $\overline{X}_\pi$ be the corresponding variety defined over the algebraic closure $\overline{\mathbb{F}}_q$ of $\mathbb{F}_q$ by the same equations as X_π. For each $r \geq 1$ there is a unique subfield $\mathbb{F}_{q^r}$ of $\overline{\mathbb{F}}_q$ such that $\mathbb{F}_{q^r}$ has q^r elements. Moreover

$$\mathbb{F}_{q^r} \subseteq \mathbb{F}_{q^t}$$

if and only if r divides t.

Let N_r be the number of points in $\overline{X}_\pi$ of the form $(x_o : \ldots : x_N)$ where each x_j lies in $\mathbb{F}_{q^r}$. Define $Z(t)$ by

$$Z(t) = \exp\left(\sum_{r \geq 1} N_r \frac{t^r}{r}\right) \in \mathbb{Q}[[t]]. \tag{10.1}$$

Example 10.1.2. If $X = \mathbb{CP}^m$, $R = \mathbb{Z}$, $\pi = p\mathbb{Z}$ then

$$N_r = 1 + p^r + p^{2r} + \ldots + p^{mr}$$

and

$$\begin{aligned} Z(t) &= \exp\left(\sum_{r \geq 1}(1 + p^r + \ldots + p^{mr})\frac{t^r}{r}\right) \\ &= \frac{1}{(1-t)(1-pt)(1-p^2t)\ldots(1-p^mt)} . \end{aligned}$$

The **Weil conjectures** relate the numbers N_r to the **Betti numbers** $\dim H_j(X)$ of X. They can be expressed in terms of the function $Z(t)$ as follows.

$$Z(t) = \frac{P_1(t)P_3(t)\ldots P_{2m-1}(t)}{P_0(t)P_2(t)\ldots P_{2m}(t)} \tag{10.2}$$

where $P_0(t) = 1 - t$, $P_{2m}(t) = 1 - q^m t$ and if $a \leq j \leq 2m - 1$ then $P_j(t)$ is a polynomial in t with integer coefficients satisfying

$$P_j(t) = \prod_{1 \leq i \leq \dim H_j(X)} (1 - \alpha_{ji}t)$$

where each α_{ji} is an algebraic integer and $|\alpha_{ji}| = q^{\frac{j}{2}}$. Note that these conditions mean that Z(t) uniquely determines the polynomials $P_j(t)$ and hence the Betti numbers of X since $\dim H_j(X) = \deg P_j(t)$.

Let $E = \sum_j(-1)^j \dim H_j(X)$ be the Euler characteristic of X. Then $Z(t)$ satisfies a functional equation

$$Z\left(\frac{1}{q^m t}\right) = \pm q^{mE/2} t^E Z(t). \tag{10.3}$$

Remark 10.1.3. The statement "$\left|\alpha_{ji}\right| - q^{\frac{j}{2}}$" is called the **Riemann hypothesis** by analogy with the Riemann Zeta function as follows. Put $t = q^{-s}$ in $Z(t)$ to get

$$Z(q^{-s}) = \exp\left(\sum_{r \geq 1} N_r \frac{q^{-rs}}{r}\right).$$

Define a prime divisor $\mathfrak{p}$ of X to be an equivalence class of points of $\overline{X}_\pi$ modulo conjugation over $\mathbb{F}_q$, and let its norm be

$$\text{Norm } \mathfrak{p} = q^{\deg \mathfrak{p}}$$

where $\deg \mathfrak{p}$ is the number of points in the equivalence class. Then since $\mathbb{F}_{q^i} \subseteq \mathbb{F}_{q^j}$ if and only if i divides j the number of points of $\overline{X}_\pi$ defined over $\mathbb{F}_{q^r}$ is

$$N_r = \sum_{\deg \mathfrak{p} | r} \deg \mathfrak{p}.$$

Hence

$$
\begin{aligned}
Z(q^{-s}) &= \exp \sum_{r \geq 1} \sum_{\deg \mathfrak{p}|r} \frac{\deg \mathfrak{p} \left(\text{Norm } \mathfrak{p}\right)^{-sr/\deg \mathfrak{p}}}{r} \\
&= \exp \sum_{\mathfrak{p}} \sum_{i} \frac{\left(\text{Norm } \mathfrak{p}\right)^{-si}}{i} \\
&= \prod_{\mathfrak{p}} \exp\left(-\log\left(1 - \left(\text{Norm } \mathfrak{p}\right)^{-s}\right)\right) \\
&= \prod_{\mathfrak{p}} \left(1 - \left(\text{Norm } \mathfrak{p}\right)^{-s}\right)^{-1}.
\end{aligned}
$$

Recall that the classical zeta function is give by

$$\zeta(s) = \sum_{n \geq 1} n^{-s} = \prod_{p \text{ prime}} (1 - p^{-s})^{-1}.$$

The classical Riemann hypothesis says that the zeros of $\zeta(s)$ lie on the line $\text{Re}(s) = \frac{1}{2}$ in $\mathbb{C}$. When $\dim_\mathbb{C} X = 1$ the statement that $\left|\alpha_{ji}\right| = q^{j/2}$ where

$$
\begin{aligned}
Z(t) &= \prod_j \left(\prod_{1 \leq i \leq \dim H_j(X)} (1 - \alpha_{ji}t)\right)^{(-1)^{j+1}} \\
&= \left(\prod_{1 \leq i \leq \dim H_1(X)} (1 - \alpha_{1i}t)\right) (1 - t)^{-1}(1 - qt)^{-1}
\end{aligned}
$$

is equivalent to the statement that if $Z(t) = 0$ then $|t| = q^{\frac{1}{2}}$, i.e. that if $Z(q^{-s}) = 0$ then $\text{Re}(s) = \frac{1}{2}$.

Weil proved some special cases of his conjecture and realised that the general case followed if one could define a suitable cohomology theory for varieties over fields of non-zero characteristic analogous to ordinary cohomology for varieties over $\mathbb{C}$. Grothendieck was able to define such a cohomology theory, ℓ-adic cohomology, using the theory of étale topology (due to himself and Artin) and thus proved part of the conjectures (the rationality of $Z(t)$ and the functional equation). Deligne finished the proof in 1973, by proving the analogue of the Riemann hypothesis. Before defining ℓ-adic cohomology let us see how its properties lead to a proof of the Weil conjectures.

10.2 Basic properties of ℓ-adic cohomology

Let Y be a quasi-projective variety over an algebraically closed field k of characteristic $p \geq 0$. Let ℓ be a prime number different from p. Let

$$\mathbb{Z}_\ell = \lim_r \mathbb{Z}/\ell^r \mathbb{Z}$$

be the ring of ℓ-adic integers, and let $\mathbb{Q}_\ell$ be its field of fractions. The ith ℓ-adic cohomology group of Y is written $H^i(Y; \mathbb{Q}_\ell)$. It has the following properties (see e.g. Milne [133]).

Proposition 10.2.1. (a) *ℓ-adic cohomology is a contravariant functor from the category of quasi-projective varieties over k to the category of vector spaces over $\mathbb{Q}_\ell$.*

(b) *$H^i(Y; \mathbb{Q}_\ell) = 0$ unless $0 \leq i \leq 2m$ where m is the dimension of Y. The dimension of $H^i(Y; \mathbb{Q}_\ell)$ is finite for all i if Y is projective (and conjecturally for any quasi-projective Y).*

(c) Poincaré duality *If Y is non-singular and projective then there is a natural perfect pairing*

$$H^i(Y; \mathbb{Q}_\ell) \otimes H^{2m-i}(Y; \mathbb{Q}_\ell) \to H^{2m}(Y; \mathbb{Q}_\ell) \cong \mathbb{Q}_\ell$$

for $0 \leq i \leq 2m$.

(d) Lefschetz fixed point formula *If Y is non-singular and projective of dimension m over k and $f: Y \to Y$ has only isolated fixed points each of multiplicity one (i.e. 1 is not an eigenvalue of the derivative of f at any $y \in Y$ such that $f(y) = y$) then the Lefschetz number $L(f)$ of f defined by*

$$L(f) = \sum_{j=0}^{2m} (-1)^j \operatorname{Tr}\left(f^*: H^j(Y; \mathbb{Q}_\ell) \to H^j(Y; \mathbb{Q}_\ell)\right)$$

is equal to the number of fixed points. More generally when f has isolated fixed points of multiplicities possibly greater than one then $L(f)$ is the number of fixed points counted according to multiplicities.

(e) **Comparison and change of base field** *If X is a non-singular complex projective variety then $H^j(X; \mathbb{Q}_\ell)$ is the ordinary cohomology of X with coefficients in $\mathbb{Q}_\ell$, so*

$$\dim_{\mathbb{C}} H^j(X) = \dim_{\mathbb{Q}_\ell} H^j(X; \mathbb{Q}_\ell).$$

Moreover if X is defined over an algebraic number ring R, as in Section 10.1, then

$$H^j(Y; \mathbb{Q}_\ell) = H^j(\overline{X}_\pi; \mathbb{Q}_\ell).$$

These are the properties of ℓ-adic cohomology which we shall need. ℓ-adic cohomology also satisfies most of the familiar properties of cohomology, such as the existence of relative cohomology, long exact sequences, spectral sequences and so on.

Let X be a non-singular complex projective variety defined over an algebraic number ring R, and define $\overline{X}_\pi$ as in Section 10.1. The properties (a)–(e) of ℓ-adic cohomology can be used to prove the Weil conjectures (10.2). The crucial ingredient is the definition of the Frobenius mapping.

Definition 10.2.2. The **Frobenius mapping** $f: \overline{X}_\pi \to \overline{X}_\pi$ is given by

$$f(x_0 : \ldots : x_N) = (x_0^q : \ldots : x_N^q).$$

This makes sense because the equations defining $\overline{X}_\pi$ as a subset of $\mathbb{P}^N(\overline{\mathbb{F}}_q)$ have coefficients in the field $\mathbb{F}_q$, and if $p(X_0, \ldots, X_N)$ is a polynomial with coefficients in $\mathbb{F}_q$ then

$$p(X_0^q, \ldots, X_N^q) = \big(p(X_0, \ldots, X_N)\big)^q.$$

A point $x \in \overline{X}_\pi$ is fixed by the rth iterate f^r of f if and only if it has coordinates in $\mathbb{F}_{q^r}$. Hence the number N_r of points in $\overline{X}_\pi$ with coordinates in $\mathbb{F}_{q^r}$ is the same as the number of fixed points of f^r. One can check that all the fixed points of f^r have multiplicity one. Thus by the Lefschetz fixed point formula (d) we have

$$N_r = L(f^r) \tag{10.4}$$

for all $r \geq 1$. This means that

$$
\begin{aligned}
Z(t) &= \exp \sum_{r \geq 1} \frac{L(f^r) t^r}{r} \\
&= \exp \sum_{r \geq 1} \sum_{0 \leq j \leq 2m} (-1)^j \mathrm{Tr}\left((f^r)^* : H^j(\overline{X}_\pi; \mathbb{Q}_\ell) \to H^j(\overline{X}_\pi; \mathbb{Q}_\ell) \right) \frac{t^r}{r} \\
&= \prod_{j=0}^{2m} \left(\exp \sum_{r \geq 1} (-1)^j \mathrm{Tr}\left((f^r)^* : H^j(\overline{X}_\pi; \mathbb{Q}_\ell) \to H^j(\overline{X}_\pi; \mathbb{Q}_\ell) \right) \frac{t^r}{r} \right) \\
&= \prod_{j=0}^{2m} \det\left(1 - t f^* : H^j(\overline{X}_\pi; \mathbb{Q}_\ell) \to H^j(\overline{X}_\pi; \mathbb{Q}_\ell) \right)^{(-1)^{j+1}}
\end{aligned}
$$

$$= \frac{P_1(t)P_3(t)\ldots P_{2m-1}(t)}{P_0(t)P_2(t)\ldots P_{2m}(t)}$$

where $P_j(t) = \det\left(1 - tf^* \colon H^j(\overline{X}_\pi, \mathbb{Q}_\ell) \to H^j(\overline{X}_\pi, \mathbb{Q}_\ell)\right)$. Then

$$P_j(t) = \prod_{1 \le i \le \dim H^j(\overline{X}_\pi, \mathbb{Q}_\ell)} (1 - \alpha_{ji}t)$$

where the α_{ji} are the eigenvalues of the action of the Frobenius map on $H^j(\overline{X}_\pi; \mathbb{Q}_\ell)$. Thus the Riemann hypothesis is equivalent to the eigenvalues of the Frobenius action on $H^j(\overline{X}_\pi; \mathbb{Q}_\ell)$ being algebraic integers of modulus $q^{j/2}$.

The functional equation (10.3) for $Z(t)$ comes straight from Poincaré duality and the fact that if $\alpha \in H^i(\overline{X}_\pi; \mathbb{Q}_\ell)$ and $\beta \in H^{2m-i}(\overline{X}_\pi; \mathbb{Q}_\ell)$ then the Poincaré pairing of $f^*\alpha$ and $f^*\beta$ is q^m times the Poincaré pairing of α and β. This is because of the naturality of the Poincaré pairing and because the Frobenius map

$$f^* \colon H^{2m}(\overline{X}_\pi; \mathbb{Q}_\ell) \to H^{2m}(\overline{X}_\pi; \mathbb{Q}_\ell)$$

is multiplication by q^m.

Having said why ℓ-adic cohomology is useful for proving the Weil conjectures, let us consider how it is defined. For more detail see Milne [133].

10.3　Étale topology and cohomology

Let Y be a quasi-projective variety defined over an algebraically closed field k. The Zariski topology on Y is the topology whose closed subsets are the subsets defined by the vanishing of homogenous polynomials (i.e. the closed subvarieties of Y). This topology reflects the algebraic structure of Y. However it is too coarse for many purposes. Of course when k is the complex field $\mathbb{C}$ we can also give Y the usual complex topology, by regarding it as a subset of a complex projective space, but in general this topology is not available.

The **étale topology** on Y plays a rôle similar to that of the complex topology. It is not a topology at all in the usual sense but it behaves in much the same way as a topology. Instead of open subsets of Y one works with étale morphisms $g \colon U \to Y$.

A morphism is étale if it is flat and unramified, or roughly speaking if it is an unbranched cover of a Zariski open subset of Y. More concretely, a morphism of varieties is étale if it induces a local isomorphism between tangent cones. In terms of algebra, if U is a quasi-projective variety over k then $g \colon U \to Y$ is an étale morphism if and only if every $u \in U$ satisfies the following condition. There are Zariski open neighbourhoods V of u in U and W of $f(u)$ in Y, functions

$$a_j \colon W \to k, \quad 1 \le j \le n,$$

such that each a_j is a rational function in inhomogeneous coordinates on W and for each $x \in W$ the polynomial

$$P(t, x) = t^n + a_1(x)t^{n-1} + \ldots + a_n(x)$$

in t has simple roots, and an isomorphism

$$V \to \{(t, x) \in k \times W \mid p(t, x) = 0\}$$

whose projection onto W is g.

If $g \colon U \to Y$ and $f \colon V \to U$ are étale morphisms then so is $g \circ f \colon V \to Y$. Moreover, if $g \colon U \to Y$ and $f \colon V \to Y$ are étale morphisms then there is a commutative diagram of étale morphisms (called a pullback diagram),

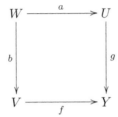

with the universal property that if

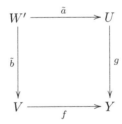

is another commutative diagram then there is a unique $\tau \colon W' \to W$ such that $\tilde{a} = a \circ \tau$ and $\tilde{b} = b \circ \tau$. The composition $g \circ a = f \circ b \colon W \to Y$ plays the rôle for the étale topology of the intersection $g \colon U \to Y$ and $f \colon V \to Y$.

The definition of a **sheaf** $\mathcal{F}$ on Y with respect to the étale topology is closely analogous to the definition of a sheaf for a genuine topology on Y.

Definition 10.3.1. For each étale morphism $g \colon U \to Y$ there is an Abelian group $\mathcal{F}(g)$ satisfying the following conditions.

(i) If $g \colon U \to Y$ and $f \colon V \to U$ are étale morphisms then there is a restriction map

$$\begin{aligned} \mathcal{F}(g) &\to \mathcal{F}(g \circ f) \\ s &\to s|_{g \circ f} \end{aligned}$$

with the usual functorial properties (Milne [133, Ch. 2 §1]).

(ii) If $g : U \to Y$ and $g_i \colon U_i \to U$ are étale morphisms such that $U = \bigcup_{i \in I} g_i(U_i)$ and if $s_i \in \mathcal{F}(g \circ g_i)$ satisfy

$$s_i|_{g \circ g_{ij}} = s_j|_{g \circ g_{ij}}$$

for all i and j where $g_{ij} \colon U_{ij} \to U$ fits into the pullback diagram

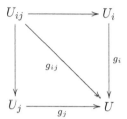

then there exists a unique $s \in \mathcal{F}(g)$ such that $s_i = s|_{g \circ g_i}$ for all $i \in I$.

Sheaf maps are defined in the obvious way and we get a category $\text{ÉtSh}(Y)$ of étale sheaves on Y.

The definition of right derived functors given in Section 3.8 for left exact additive functors between categories of sheaves can be adapted directly to define the right derived functors of functors from $\text{ÉtSh}(Y)$ to Ab. There is a functor $\Gamma_Y \colon \text{ÉtSh}(Y) \to \text{Ab}$ defined by

$$\Gamma_Y(\mathcal{F}) = \mathcal{F}(1_Y \colon Y \to Y) \tag{10.5}$$

where 1_Y is the identity map on Y and Γ_Y is a left exact additive functor. The étale cohomology groups of $\mathcal{F}$ are defined to be the right derived functors of Γ_Y applied to $\mathcal{F}$:

$$H^i_{\text{ét}}(Y; \mathcal{F}) = R^i\Gamma_Y(\mathcal{F}). \tag{10.6}$$

Remark 10.3.2. Alternatively one can adapt the definition of Čech cohomology and define étale Čech cohomology groups

$$\check{H}^i_{\text{ét}}(Y; \mathcal{F})$$

which for sufficiently well-behaved sheaves $\mathcal{F}$ are canonically isomorphic to the groups

$$H^i_{\text{ét}}(Y; \mathcal{F}).$$

Now suppose that ℓ is any prime number different from the characteristic p of k. The constant sheaf $(\mathbb{Z}/\ell^r\mathbb{Z})_Y$ on Y is defined by

$$(\mathbb{Z}/\ell^r\mathbb{Z})_Y(g \colon U \to Y) = \{\text{continuous maps } h \colon U \to \mathbb{Z}/\ell^r\mathbb{Z}\}$$

where U is given the Zariski topology and $\mathbb{Z}/\ell^r\mathbb{Z}$ has the discrete topology (so that a continuous map $h \colon U \to \mathbb{Z}/\ell^r\mathbb{Z}$ is constant on every connected component of U). The restriction map

$$(\mathbb{Z}/\ell^r\mathbb{Z})_Y(g) \to (\mathbb{Z}/\ell^r\mathbb{Z})_Y(g \circ f)$$

is given by composition with f.

Definition 10.3.3. The ℓ-adic cohomology of Y is defined to be

$$H^*(Y;\mathbb{Q}_\ell) = (\lim H^*_{\text{ét}}(Y;(\mathbb{Z}/\ell^r\mathbb{Z})_Y)) \otimes \mathbb{Q}_\ell.$$

Remark 10.3.4. This definition is rather subtle; the ℓ-adic cohomology is *not* simply the étale cohomology of a constant sheaf. We must first take the étale cohomology with coefficients in $(\mathbb{Z}/\ell^r\mathbb{Z})_Y$ and only then pass to the limit and tensor with the field of fractions. One way to think of this is that we must work with categories of projective systems and avoid passing to the limit (even though it exists) too soon.

10.4 The Weil conjectures for singular varieties and ℓ-adic intersection cohomology

Suppose that $Y = \overline{X}_\pi$ where $\overline{X}_\pi$ is the reduction modulo a prime ideal π of a complex projective variety $X \subseteq \mathbb{CP}^N$ defined over an algebraic number ring R and $\overline{X}_\pi$ is the extension of X to a variety defined over the algebraic closure $\overline{\mathbb{F}}_q$ of $\mathbb{F}_q = R/\pi$. We have seen that if X is non-singular and π is chosen appropriately then the properties of the ℓ-adic cohomology of Y can be used to prove the Weil conjectures for X. What happens when X is allowed to be singular? The Weil conjectures certainly fail as they are stated, but they can be made to work if one uses intersection cohomology throughout instead of ordinary cohomology. To see why this might be true we must define ℓ-adic intersection cohomology.

By enlarging the algebraic number ring R if necessary, we can assume that X has a Whitney stratification given by a filtration

$$X = X_m \supseteq X_{m-1} \supseteq \cdots \supseteq X_0$$

where each X_j is defined over R. Moreover we can assume that if Y_j is the extension to $\overline{\mathbb{F}}_q$ of the reduction of X_j modulo π then

$$Y = Y_m \supseteq Y_{m-1} \supseteq \cdots \supseteq Y_0$$

is a filtration of

$$Y = \overline{X}_\pi$$

by closed subvarieties Y_j such that $Y_j - Y_{j-1}$ is either empty or is non-singular of dimension j for each j. We can now use Deligne's construction of intersection homology (Section 7.3) to define the ℓ-adic intersection cohomology

$$IH^*(Y;\mathbb{Q}_\ell)$$

of Y as follows. Let $i_k : Y - Y_{m-k} \hookrightarrow Y - Y_{m-k-1}$ be the inclusion. Define a complex of sheaves $\mathcal{IC}_Y^\bullet(\mathbb{Z}/\ell^r\mathbb{Z})$ on Y in the étale topology by

$$\tau_{\leq -m-1}R(i_m)_* \cdots \tau_{\leq -2m+1}R(i_2)_*\tau_{\leq -2m}R(i_1)_*(\mathbb{Z}/\ell^r\mathbb{Z})_{Y-Y_{m-1}}[2m] \quad (10.7)$$

where

$$(\mathbb{Z}/\ell^r\mathbb{Z})_{Y-Y_{m-1}}[2m]$$

is the complex on $Y - Y_{m-1}$ which is the constant sheaf $(\mathbb{Z}/\ell^r\mathbb{Z})_{Y-Y_{m-1}}$ in degree $i = -2m$ and 0 in other degrees (all with respect to the étale topology). By analogy with the ℓ-adic cohomology we define

$$IH_*(Y;\mathbb{Q}_\ell) = (\lim H^*_{\text{ét}}(Y;\mathcal{IC}^\bullet_Y(\mathbb{Z}/\ell^r\mathbb{Z}))) \otimes \mathbb{Q}_\ell$$

and let the ℓ-adic intersection cohomology $IH^*(Y;\mathbb{Q}_\ell)$ be its dual. For more details see Beilinson, Bernstein and Deligne [13], Brylinski [36].

Proposition 10.4.1. *For a (possibly singular) complex projective variety X defined over an algebraic number ring R the ℓ-adic intersection cohomology thus defined has the following properties.*

(i) **Comparison and change of base field.**
   With the notation above we have

$$IH^*(Y;\mathbb{Q}_\ell) \cong IH^*(X;\mathbb{Q}_\ell)$$

   where $Y = \overline{X}_\pi$. Moreover

$$\dim_{\mathbb{Q}_\ell} IH^i(X;\mathbb{Q}_\ell) = \dim_{\mathbb{C}} IH^i(X).$$

(ii) **Poincaré duality.**
   There is a perfect pairing

$$IH^i(Y;\mathbb{Q}_\ell) \otimes IH^{2m-i}(Y;\mathbb{Q}_\ell) \to \mathbb{Q}_\ell.$$

(iii) **Lefschetz fixed point formula.**
   If $f\colon Y \to Y$ is an isomorphism with isolated fixed points then the Lefschetz number

$$L(f) = \sum_{j=0}^{2m}(-1)^j\,\text{Tr}\left(f^*\colon IH^j(Y;\mathbb{Q}_\ell) \to IH^j(Y;\mathbb{Q}_\ell)\right)$$

   of f is equal to the number of fixed points of f counted according to multiplicity. Unfortunately the definition of multiplicity becomes more complicated when the fixed point is a singularity of Y (cf. Goresky and MacPherson [72]).

As in Section 10.2 we consider the Frobenius map $f\colon \overline{X}_\pi \to \overline{X}_\pi$ defined by

$$f(x_0 : \ldots : x_N) = (x_0^q : \ldots : x_N^q).$$

The Lefschetz number $L(f^r)$ of the rth iterate of f is the number of points of $\overline{X}_\pi$ defined over the field $\mathbb{F}_{q^r}$, but counted according to multiplicity. The multiplicity of a non-singular point is one, but in general the multiplicity depends on the singularity of $\overline{X}_\pi$ at the point in question.

Just as in the non-singular case, the Frobenius map acts trivially on $IH^0(\overline{X}_\pi; \mathbb{Q}_\ell)$ and as multiplication by q^m on $IH^{2m}(\overline{X}_\pi; \mathbb{Q}_\ell)$. Moreover the eigenvalues of its action on

$$IH^j(\overline{X}_\pi; \mathbb{Q}_\ell)$$

for any j between 0 and $2m$ are algebraic integers α_{ji} with modulus

$$|\alpha_{ji}| = q^{\frac{i}{2}}. \tag{10.8}$$

This fact, sometimes called the Riemann hypothesis as in Section 10.1, is very important. (Its proof makes use of Poincaré duality: once it has been shown that $|\alpha_{ji}| \leq q^{\frac{i}{2}}$ for all j then Poincaré duality gives the reverse inequality $|\alpha_{ji}| \geq q^{\frac{i}{2}}$.) Its importance is not merely that it can be used to generalise the Weil conjectures to apply to singular projective varieties, provided one uses intersection homology and counts points according to multiplicities depending on the singularities of the points. Its main importance is that in any reasonable cohomology theory such as ℓ-adic intersection cohomology there are natural boundary maps and degeneracy maps appearing in long exact sequences, spectral sequences, etc. These maps often go from the cohomology of one space to the cohomology of another space in a *different* dimension. Because these maps are natural, in the case of the ℓ-adic intersection cohomology groups of subvarieties of $\mathbb{P}^N(\overline{\mathbb{F}}_q)$ defined over $\mathbb{F}_q$ they must commute with the Frobenius maps. But since the eigenvalues of the Frobenius maps acting on ℓ-adic intersection cohomology groups of projective varieties in different dimensions are different this means that the boundary and degeneracy maps between such intersection cohomology groups must vanish. Using the comparison theorem one finds that the corresponding boundary maps and degeneracy maps for the ordinary intersection cohomology of complex projective varieties must vanish also. This enables one to prove theorems about ordinary intersection cohomology of complex varieties. An important example is Beilinson, Bernstein, Deligne and Gabber's decomposition theorem (see Chapter 8).

Remark 10.4.2. At first sight this argument only applies when the complex varieties involved are defined over algebraic number rings. However a finite set of equations defining any complex projective variety can be "deformed" slightly without altering the intersection cohomology so that the equations become equations with coefficients in an algebraic number field.

10.5   Further reading

Serre [161] is a very lucid survey of zeta and L-functions. Katz [103] gives an overview of Deligne's proof of the Weil conjectures. A much more detailed, and fairly self-contained, account can be found in Freitag and Kiehl [59] and Kiehl and Weissauer [109]. The former deals with étale cohomology and Deligne's proof of the Weil conjecture in the non-singular case, the latter with perverse sheaves on schemes and the extension of the proof to singular varieties.

Chapter 11

$\mathcal{D}$-Modules and the Riemann–Hilbert correspondence

This chapter contains a brief sketch of the theory of $\mathcal{D}$-modules and their relationship to intersection homology.

11.1 The Riemann–Hilbert problem

Consider the system of m first-order differential equations

$$\frac{df_i}{dz} = \sum_{j=1}^{m} a_{ij}(z)f_j(z), \quad 1 \le i \le m \tag{11.1}$$

in m complex-valued functions of one complex variable z, where each $a_{ij}(z)$ is a meromorphic function of z defined on a connected open subset U of $\mathbb{CP}^1 = \mathbb{C} \cup \{\infty\}$.

Example 11.1.1. A single mth order differential equation

$$\frac{d^m f}{dz^m} + a_1(z)\frac{d^{m-1}f}{dz^{m-1}} + \ldots + a_m(z)f(z) = 0$$

is equivalent to the system of equations

$$\frac{df_i}{dz} = f_{i+1}, \quad 1 \le i \le m-1,$$

$$\frac{df_m}{dz} = -a_1(z)f_m(z) - \ldots - a_m(z)f_1(z).$$

If each meromorphic function $a_{ij}(z)$ is holomorphic on U then the solutions of the system (11.1) are multi-valued holomorphic functions of $z \in U$ and the space Σ of solutions is a vector space of dimension m. However if at least one of the coefficients $a_{ij}(z)$ has a singularity at some $b \in U$ then in general the solutions have branch points at b and b is called a **singular point** of the system.

Definition 11.1.2. A singular point $b \in U$ is called a **regular singular point** of the system (11.1) if whenever

$$(f_1(z), \ldots, f_m(z))$$

is a multi-valued solution of the system near b then there is some positive integer r such that

$$|r - b|^r f_j(z) \to 0$$

for each j as $z \to b$. Otherwise b is called an **irregular singular point**.

When all functions $a_{ij}(z)$ are rational (i.e. they are meromorphic on $\mathbb{CP}^1$) then the system (11.1) is said to be of **Fuchsian type** if all the singular points are regular. The mth order equation in Example 11.1.1 is said to be of Fuchsian type if the corresponding system of m first order differential equations is of Fuchsian type.

Example 11.1.3. Let α be a fixed complex number. The equation

$$\frac{df}{dz} = \frac{\alpha}{z} f(z)$$

has solutions $f(z) = cz^\alpha$ for $c \in \mathbb{C}$. If α is not an element of $\mathbb{Z}$ these solutions are multi-valued with branch points at 0, and 0 is a regular singular point of the system.

In fact the system (11.1) has a regular singular point at 0 if and only if it is equivalent to a system of the same form such that the coefficients $a_{ij}(z)$ have poles of order at most one at 0 (see e.g. Borel [20, III, 1.3.1]).

Now let $b_0, \ldots, b_k$ be the points of U which are singular points for the system (11.1). If γ is a closed path in

$$U - \{b_0, \ldots, b_k\}$$

then analytic continuation along γ induces a linear transformation $\phi(\gamma) : \Sigma \to \Sigma$ of the space of solutions. If we choose a basis of Σ we get a representation

$$\phi : \pi_1(U - \{b_0, \ldots, b_k\}) \to GL(m, \mathbb{C}) \tag{11.2}$$

of the fundamental group of $U - \{b_0, \ldots, b_k\}$. This representation ϕ is called the **monodromy** of the system (with respect to the chosen basis of Σ). Note that up to a choice of basis such a representation ϕ corresponds exactly to a

local system $\mathcal{L}$ on $U - \{b_0, \ldots, b_k\}$ with $\mathcal{L}_x \cong \mathbb{C}^m$ for all $x \in U - \{b_0, \ldots, b_k\}$ (cf. Section 4.9).

In 1857 Riemann posed the following problem. Given points $b_0, \ldots, b_k \in \mathbb{CP}^1$ and a faithful representation

$$\phi : \pi_1(U - \{b_0, \ldots, b_k\}) \to \mathrm{GL}(m, \mathbb{C}) \tag{11.3}$$

find all systems of Fuchsian type whose singular points are $b_0, \ldots, b_k$ and whose monodromy (with respect to some basis of the space of solutions) is ϕ. Riemann showed that when $m = k = 2$ there is a unique system of Fuchsian type with given singular points b_0, b_1, b_2 and given monodromy

$$\phi : \pi_1(\mathbb{CP}^1 - \{b_0, b_1, b_2\}) \to GL(2; \mathbb{C}).$$

When the singular points are $0, 1, \infty$ this system is given by the hypergeometric equation

$$z(z - 1)\frac{d^2 f}{dz^2} + (\gamma - (\alpha + \beta + 1)z)\frac{df}{dz} - \alpha\beta f = 0 \tag{11.4}$$

where α, β, γ are constants depending on the monodromy ϕ.

When in 1900 Hilbert listed twenty three problems as targets for mathematicians in the twentieth century he included a generalisation of Riemann's question. It is easy to extend to arbitrary compact Riemann surfaces the definitions of systems of first order differential equations with meromorphic coefficients, systems of Fuchsian type and monodromy. (One way is to identify functions on a fixed compact Riemann surface S with multi-valued functions on $\mathbb{CP}^1$.) Suppose we are given a compact Riemann surface S, points $b_0, \ldots, b_k$ of S and a representation

$$\phi : \pi_1(S - \{b_0, \ldots, b_k\}) \to \mathrm{GL}(m, \mathbb{C}) \tag{11.5}$$

of the fundamental group $\pi_1(S - \{b_0, \ldots, b_k\})$. Hilbert's twenty first problem (often called the **Riemann–Hilbert problem**) was to find those systems of differential equations of Fuchsian type over S whose monodromy is ϕ.

Many mathematicians worked on this problem and it was finally shown a hundred years after Riemann posed his original question that there is an exact correspondence between systems of Fuchsian type and their monodromy representations (see Röhrl [147]). This correspondence is often called the **Riemann–Hilbert correspondence**. However systems with irregular singularities are not determined by their monodromy representations.

So far we have been considering systems of differential equations in *one* complex variable, i.e. over a 1-dimensional complex manifold. In this chapter we shall discuss a more general form of the Riemann–Hilbert correspondence which relates differential systems (or $\mathcal{D}$-modules) on a complex quasi-projective variety X to the intersection sheaf complexes of subvarieties of X with coefficients in local systems.

11.2   Differential systems over $\mathbb{C}^n$

Fix $n \geq 1$ and let $\mathcal{O}$ denote either the ring of holomorphic functions on $\mathbb{C}^n$ or the ring of polynomial functions on $\mathbb{C}^n$. The choice we make depends on whether we wish later to study holomorphic $\mathcal{D}$-modules or algebraic $\mathcal{D}$-modules. We shall mainly be interested in algebraic $\mathcal{D}$-modules, but the theories are very closely related. Let $\mathcal{D}$ be the ring of differential operators generated by the ring $\mathcal{O}$ together with

$$D_1, D_2, \ldots, D_n$$

(which are to be thought of as $\frac{\partial}{\partial z_1}, \ldots, \frac{\partial}{\partial z_n}$ where $z_1, \ldots, z_n$ are complex coordinates on $\mathbb{C}^n$), satisfying the relations

$$D_i D_j = D_j D_i,$$

$$D_i g = g D_i + \frac{\partial g}{\partial z_i} \quad \text{if } g \in 0.$$

Then $\mathcal{D}$ acts on $\mathcal{O}$ via $g \cdot f = gf$, $D_i \cdot f = \frac{\partial f}{\partial z_i}$ for $f \in \mathcal{O}$.

Definition 11.2.1. A **differential system** on $\mathbb{C}^n$ is a left $\mathcal{D}$-module $\mathcal{M}$ such that there is an exact sequence of left $\mathcal{D}$-modules

$$\mathcal{D}^p \to \mathcal{D}^q \to \mathcal{M} \to 0$$

where $\mathcal{D}^p = \mathcal{D} \oplus \ldots \oplus \mathcal{D}$ is the direct sum of $\mathcal{D}$ with itself p times. A **solution** of the differential system $\mathcal{M}$ with values in a left $\mathcal{D}$-module $\mathcal{F}$ is a map of left $\mathcal{D}$-modules

$$\phi : \mathcal{M} \to \mathcal{F}.$$

Motivation

If $\mathcal{M}$ is a differential system with an exact sequence

$$\mathcal{D}^p \to \mathcal{D}^q \to \mathcal{M} \to 0$$

then $\mathcal{M}$ is generated as a left $\mathcal{D}$-module by the images $f_1, \ldots, f_q$ under the surjection $\mathcal{D}^q \to \mathcal{M}$ of the usual basis $e_1, \ldots, e_q$ of $\mathcal{D}^q$ as left $\mathcal{D}$-module. Moreover the kernel of this surjection is the image of the map $\mathcal{D}^p \to \mathcal{D}^q$. Hence it is generated as a left $\mathcal{D}$-module by the images $r_1, \ldots, r_p$, say, of the standard basis of $\mathcal{D}^p$. We can write

$$r_i = \sum_{1 \leq j \leq q} d_{ij} e_j$$

where each d_{ij} is an element of $\mathcal{D}$. Then the generators $f_1, \ldots, f_q$ of $\mathcal{M}$ satisfying the relations

$$\sum_{1 \leq j \leq q} d_{ij} f_j = 0, \qquad 1 \leq i \leq p.$$

A solution ϕ of $\mathcal{M}$ with values in $\mathcal{F}$ is uniquely determined by the images $\phi(f_1), \dots, \phi(f_q)$ of generators $f_1, \dots, f_q$ of $\mathcal{M}$ in $\mathcal{F}$. If $\phi_1, \dots, \phi_q$ are elements of $\mathcal{F}$ then there is a solution $\phi : \mathcal{M} \to \mathcal{F}$ such that $\phi(f_j) = \phi_j$ for $1 \leq j \leq q$ if and only if the ϕ_j satisfy the equations

$$\sum_{1 \leq j \leq q} d_{ij} \phi_j = 0, \qquad 1 \leq i \leq p.$$

Thus a differential system $\mathcal{M}$ on $\mathbb{C}^n$ together with a choice of generators and relations for $\mathcal{M}$ is 'equivalent' to a finite set of partial differential equations in a finite number of unknown functions in the variables $z_1, \dots, z_n$.

Example 11.2.2.   1. The equations

$$\frac{\partial f}{\partial z_1} = 0, \quad z_1 \frac{\partial f}{\partial z_2} + z_2 \frac{\partial f}{\partial z_3} + \dots + z_{n-1} \frac{\partial f}{\partial z_n} = 0$$

define a differential system $\mathcal{M}_1$ on $\mathbb{C}^n$ with one generator f and two relations $D_1 f$ and $(z_1 D_2 + z_2 D_3 + \dots + z_{n-1} D_n) f$. Thus there is an exact sequence of left $\mathcal{D}$-modules

$$\mathcal{D}^2 \to \mathcal{D} \to \mathcal{M}_1 \to 0.$$

2. The equations

$$\frac{\partial f}{\partial z_1} = 0, \quad \frac{\partial f}{\partial z_2} = 0, \quad \dots \quad , \frac{\partial f}{\partial z_n} = 0$$

define a differential system $\mathcal{M}_2$ on $\mathbb{C}^n$ with one generator f and n relations $D_1 f, \dots, D_n f$. Recall that the **commutator** of two elements δ_1, δ_2 of $\mathcal{D}$ is

$$[\delta_1, \delta_2] = \delta_1 \delta_2 - \delta_2 \delta_1.$$

The differential systems $\mathcal{M}_1$ and $\mathcal{M}_2$ are isomorphic (as left $\mathcal{D}$-modules) because

$$[D_j, z_1 D_2 + z_2 D_3 + \dots + z_{n-1} D_n] = D_{j+1}$$

for $1 \leq j < n$, so the left ideal of $\mathcal{D}$ generated by D_1 and $z_1 D_2 + \dots + z_{n-1} D_n$ is the same as the left ideal generated by $D_1, \dots, D_n$.

3. When $n = 1$ write z for z_1 and D for D_1.

Consider the differential system $\mathcal{M}_3$ with one generator f and one relation

$$D^m f + a_1 D^{m-1} f + \dots + a_m f$$

where $a_1, \dots, a_m \in \mathcal{O}$, corresponding to the differential equation

$$\frac{\partial^m f}{\partial z^m} + a_1(z) \frac{\partial^{m-1} f}{\partial z^{m-1}} + \dots + a_m(z) f = 0.$$

It was observed in Example 11.1.1 that $\mathcal{M}_3$ is isomorphic to the differential system $\mathcal{M}_4$ with m generators $f_1, \ldots, f_m$ and m relations

$$Df_1 - f_2, \ldots, Df_{m-1} - f_m$$

and

$$Df_m + a_1 f_m + \ldots + a_m f_1.$$

11.3 $\mathcal{D}_X$-modules and intersection homology

We can globalise the definition of a differential system. Let X be either a complex manifold or a complex quasi-projective variety. Denote by $\mathcal{O}_X$ the sheaf of holomorphic (respectively regular) functions on X. That is, if U is an open subset of X (in either the complex topology or the Zariski topology) then

$$\mathcal{O}_X(U) = \{\text{holomorphic functions } h : U \to \mathbb{C}\}$$

or

$$\mathcal{O}_X(U) = \{\text{regular functions } h : U \to \mathbb{C}\}.$$

Recall that a regular function h is one which can be expressed locally with respect to homogeneous coordinates on the ambient projective space as the quotient P/Q of a homogeneous polynomial P by a locally non-vanishing homogeneous polynomial Q of the same degree. A regular function $\mathbb{C}^n \to \mathbb{C}$ is just a polynomial function.

A **differential operator** on X is a sheaf map

$$\delta : \mathcal{O}_X \to \mathcal{O}_X$$

such that in local coordinates δ is given by a differential operator on a subset of $\mathbb{C}^n$ with either holomorphic or regular coefficients. Equivalently for some positive integer k and every open $U \subseteq X$

$$\delta(U) : \mathcal{O}_X(U) \to \mathcal{O}_X(U)$$

satisfies

$$[\hat{f}_k, [\hat{f}_{k-1} \ldots [\hat{f}_1, [\hat{f}_0, \delta]] \ldots]] = 0$$

for any $f_0, \ldots, f_k \in \mathcal{O}_X(U)$ where $\hat{f}_j$ is the operator on $\mathcal{O}_X(U)$ given by multiplication by f_j.

The sheaf $\mathcal{D}_X$ on X is defined by

$$\mathcal{D}_X(U) = \{\text{differential operators on } U\}$$

for U open in X. Note that both $\mathcal{O}_X$ and $\mathcal{D}_X$ are sheaves of rings on X, in the sense that if U is open in X then $\mathcal{O}_X(U)$ and $\mathcal{D}_X(U)$ are rings and the restriction maps preserve the ring structure.

If $\mathcal{A}$ is a sheaf of rings on X then a left $\mathcal{A}$-module is sheaf $\mathcal{M}$ on X such that for each open U in X the Abelian group $\mathcal{M}(U)$ is an $\mathcal{A}(U)$-module and the restriction maps respect the module structure.

Definition 11.3.1. A sheaf of rings $\mathcal{A}$ on X is called **coherent** if given any map of left $\mathcal{A}$-modules

$$\theta : \mathcal{A}^p \to \mathcal{A}^q$$

then for all $x \in X$ there exist open neighbourhoods U of x in X and finitely many sections $\sigma_1, \ldots, \sigma_r$ of $\ker \theta$ over U such that $\sigma_1, \ldots, \sigma_r$ generate $\ker \theta \big|_U$ as an $\mathcal{A} \big|_U$-module. That is, the map

$$(\mathcal{A} \big|_U)^r \to \ker \theta \big|_U$$

given by sending $(\alpha_1, \ldots, \alpha_r) \in \mathcal{A}(V)^r$ to

$$\alpha_1 \sigma_1 \big|_V + \ldots + \alpha_r \sigma_r \big|_V \in \ker \theta(V)$$

for $V \subseteq U$ is surjective. Equivalently there is an exact sequence of left $\mathcal{A} \big|_U$-modules

$$\mathcal{A}^r \big|_U \to \mathcal{A}^p \big|_U \xrightarrow{\theta} \mathcal{A}^q \big|_U.$$

Theorem 11.3.2. *(see e.g. Borel [20, II §3]). $\mathcal{O}_X$ and $\mathcal{D}_X$ are coherent sheaves of rings on X.*

Definition 11.3.3. If $\mathcal{A}$ is a coherent sheaf of rings on X then a left $\mathcal{A}$-module $\mathcal{M}$ is called **coherent** if every $x \in X$ has an open neighbourhood U in X such that there is an exact sequence

$$\mathcal{A}^p \big|_U \to \mathcal{A}^q \big|_U \to \mathcal{M} \big|_U \to 0$$

of left $\mathcal{A} \big|_U$-modules.

Coherent $\mathcal{A}$-modules on X are better behaved than $\mathcal{A}$-modules in general. Some pathological examples are avoided by imposing the condition of coherence (cf. Pham [143, §2.6]).

The natural way to globalise the definition of a differential system is now the following.

Definition 11.3.4. A **differential system** on X is a coherent $\mathcal{D}_X$-module $\mathcal{M}$.

We shall see that differential systems are closely related to intersection homology. Let X be a non-singular complex projective variety of dimension n. Let Ω_X^r be the sheaf of holomorphic sections of the bundle $\Lambda^r T^* X$ where $T^* X$ is the *complex* cotangent bundle of X. Then Ω_X^r is a left $\mathcal{O}_X$-module. A local section ω of Ω_X^r is given in local coordinates $z_1, \ldots, z_n$ on X by

$$w(z) = \sum_{i_1 < \ldots < i_r} a_{i_1 \ldots i_r}(z) dz_{i_1} \wedge \ldots \wedge dz_{i_r}$$

where the coefficients $a_{i_1 \ldots i_r}$ are holomorphic functions of $z = (z_1, \ldots, z_n)$. We define $d : \Omega_X^r \to \Omega_X^{r+1}$ in local coordinates by

$$dw(z) = \sum_{i_1 < \ldots < i_r} \sum_{1 \le k \le n} \left(\frac{\partial a_{i_1 \ldots i_r}}{\partial z_k} \right) dz_k \wedge dz_{i_1} \wedge \ldots \wedge dz_{i_r}.$$

Given a coherent $\mathcal{D}_X$-module $\mathcal{M}$ we define the **de Rham complex** of $\mathcal{M}$ to be the complex $\mathrm{DR}(\mathcal{M})$ given by

$$0 \longrightarrow \mathcal{M} \xrightarrow{d_{\mathcal{M}}} \Omega_X^1 \otimes_{\mathcal{O}_X} \mathcal{M} \xrightarrow{d_{\mathcal{M}}} \Omega_X^2 \otimes_{\mathcal{O}_X} \mathcal{M} \cdots \Omega_X^n \otimes_{\mathcal{O}_X} \mathcal{M} \longrightarrow 0$$

where in local coordinates $z_1, \ldots, z_n$ the sheaf map $d_{\mathcal{M}}$ is given by

$$d_{\mathcal{M}}(w \otimes m) = dw \otimes m + \sum_{1 \le k \le n} (dz_k \wedge w) \otimes D_k w.$$

(cf. Pham [143, §2.14.2]).

The **Riemann–Hilbert correspondence** will tell us that, under the de Rham functor DR, irreducible holonomic $\mathcal{D}_X$-modules with regular singularities correspond exactly to the intersection sheaf complexes of irreducible subvarieties of X with coefficients in local systems, up to generalised quasi-isomorphism.

In order to explain the Riemann–Hilbert correspondence we must define holonomic $\mathcal{D}_X$-modules with regular singularities.

11.4 The characteristic variety of a $\mathcal{D}_X$-module

Let $P \in \mathcal{D}$ be a differential operator on $\mathbb{C}^n$. Then we can write

$$P = \sum_{|\alpha| \le m} c_\alpha(z) D^\alpha$$

where $\alpha = (\alpha_1, \ldots, \alpha_n) \in \mathbb{N}^n$, $|\alpha| = \alpha_1 + \ldots + \alpha_n$, $D^\alpha = D_1^{\alpha_1} \ldots D_n^{\alpha_n}$ and $c_\alpha(z) \in \mathcal{O}$. If m is chosen as small as possible then m is called the **order** of P, and the **principal symbol** of P is

$$\sigma(P) = \sum_{|\alpha| = m} c_\alpha(z) \xi^\alpha \in \mathcal{O}[\xi_1, \ldots, \xi_n]$$

where $\xi^\alpha = \xi_1^{\alpha_1} \ldots \xi_n^{\alpha_n}$ and $\mathcal{O}[\xi_1, \ldots, \xi_n]$ is the polynomial ring in $\xi_1, \ldots, \xi_n$ with coefficients in $\mathcal{O}$. For any $m \in \mathbb{N}$ we define the m**th symbol** $\sigma_m(P)$ of P by the same formula.

Let $\mathcal{D}^{(m)} \subseteq \mathcal{D}$ be the set of differential operators of order at most m. (By convention the operator 0 has order $-\infty$). Let Σ be the ring

$$\Sigma = \bigoplus_{m \ge 0} \mathcal{D}^{(m)} / \mathcal{D}^{(m-1)}$$

with multiplication

$$\left(\mathcal{D}^{(m)}/\mathcal{D}^{(m-1)}\right) \otimes \left(\mathcal{D}^{(\ell)}/\mathcal{D}^{(\ell-1)}\right) \to \mathcal{D}^{(\ell+m)}/\mathcal{D}^{(\ell+m-1)}$$

defined by the composition of differential operators

$$\mathcal{D}^{(m)} \otimes \mathcal{D}^{(\ell)} \to \mathcal{D}^{(\ell+m)}.$$

There is an isomorphism

$$\Sigma \to \mathcal{O}[\xi_1, \ldots, \xi_n]$$

whose restriction to $\mathcal{D}^{(m)}/\mathcal{D}^{(m-1)}$ is induced by the symbol σ_m. We shall use this isomorphism to identify Σ with the polynomial ring $\mathcal{O}[\xi_1, \ldots, \xi_n]$.

Now consider a differential system $\mathcal{M}$ on $\mathbb{C}^n$ with a given exact sequence

$$\mathcal{D}^p \to \mathcal{D}^q \to \mathcal{M} \to 0.$$

Let $\mathcal{M}^{(m)}$ be the image of $\left(\mathcal{D}^{(m)}\right)^q$ in $\mathcal{M}$, and let

$$\mathrm{Gr}\mathcal{M} = \bigoplus_{m \geq 0} \frac{\mathcal{M}^{(m)}}{\mathcal{M}^{(m-1)}}.$$

Then $\mathrm{Gr}\mathcal{M}$ is a coherent Σ-module. Let $\mathcal{I}$ be the ideal in Σ which is the annihilator of $\mathrm{Gr}\mathcal{M}$ and let $\sqrt{\mathcal{I}}$ be the radical. Then

$$\mathcal{I} = \{p \in \mathcal{O}[\xi_1, \ldots, \xi_n] \mid pu = 0, \quad \forall u \in \mathrm{Gr}\mathcal{M}\}$$

and

$$\sqrt{\mathcal{I}} = \{p \in \mathcal{O}[\xi_1, \ldots, \xi_n] \mid \exists k \geq 1, p^k \in \mathcal{I}\}.$$

Theorem 11.4.1. *$\sqrt{\mathcal{I}}$ depends only on $\mathcal{M}$, not on the choice of exact sequence*

$$\mathcal{D}^p \to \mathcal{D}^q \to \mathcal{M} \to 0.$$

Sketch of proof. (For more details see e.g. Pham [143, §2.8].) One shows that $\sqrt{\mathcal{I}}$ is a homogeneous ideal (that is, it is generated by homogeneous polynomials) and that the following two statements hold.

(i) If $P \in \mathcal{D}^{(m)}$ and $p = \sigma_m(P)$ then $p \in \sqrt{\mathcal{I}}$ if and only if

$$p^s \mathcal{M}^{(\ell)} \subseteq \mathcal{M}^{(\ell+ms-r(s))} \quad \forall \ell \in \mathbb{N}, \quad s \in \mathbb{N}$$

where $r(s) \to \infty$ as $s \to \infty$.

(ii) If $\{\widetilde{\mathcal{M}}^{(m)} \mid m \geq 0\}$ is another such filtration of $\mathcal{M}$ then there exist $\lambda, \mu \in \mathbb{N}$ such that

$$\mathcal{M}^{(\ell)} \subseteq \widetilde{\mathcal{M}}^{(\ell+\lambda)}, \quad \widetilde{\mathcal{M}}^{(\ell)} \subseteq \mathcal{M}^{(\ell+\mu)}$$

for all $\ell \geq 0$.

This is enough to prove the theorem.

We can globalise Theorem 11.4.1. Let X be a complex manifold or a non-singular quasi-projective complex variety as before. Then the sheaf of rings $\mathcal{D}_X$ has a filtration by the $\mathcal{D}_X^{(m)}$ in the obvious way and the associated graded sheaf of rings

$$\mathrm{Gr}\mathcal{D}_X = \bigoplus_{m \geq 0} \frac{\mathcal{D}_X^{(m)}}{\mathcal{D}_X^{(m-1)}}$$

can be naturally identified with the sheaf of holomorphic (or regular) functions on T^*X which are polynomial in the variables in the fibre direction.

If U is an open subset of X and

$$P \in \mathcal{D}_X^{(m)}(U)$$

is a differential operator of order m over U then the symbol $\sigma(P)$ of P is the image of P under the composition

$$\mathcal{D}_X^{(m)}(U) \to \frac{\mathcal{D}_X^{(m)}(U)}{\mathcal{D}_X^{(m-1)}(U)} \to \mathrm{Gr}\mathcal{D}_X(U).$$

Now let $\mathcal{M}$ be a coherent $\mathcal{D}_X$-module. One can always find a 'good filtration' *locally* (Borel [20, II §4], Pham [143, §2.8])

$$\mathcal{M}^{(0)} \subseteq \mathcal{M}^{(1)} \subseteq \ldots \subseteq \mathcal{M}^{(m)} \subseteq \ldots$$

of $\mathcal{M}$; that is, a filtration which satisfies the conditions

(i) $\mathcal{D}^{(r)}\mathcal{M}^{(m)} \subseteq \mathcal{M}^{(r+m)}$ with equality when m is sufficiently large;

(ii) $\mathcal{M}_{(m)}$ is a coherent $\mathcal{O}_X$-module.

Then (locally)

$$\mathrm{Gr}\mathcal{M} = \bigoplus_{m \geq 0} \frac{\mathcal{M}^{(m)}}{\mathcal{M}^{(m-1)}}$$

is a coherent sheaf of $\mathrm{Gr}\mathcal{D}_X$-modules, and its annihilator $\mathcal{I}$ is a coherent sheaf of ideals in $\mathrm{Gr}\mathcal{D}_X$. Moreover locally $\sqrt{\mathcal{I}}$ is a coherent sheaf of ideals in $\mathrm{Gr}\mathcal{D}_X$ which is *independent* of the choice of filtration. This means that $\sqrt{\mathcal{I}}$ is well defined globally as a sheaf of ideals in $\mathrm{Gr}\mathcal{D}_X$, which is a sheaf of functions on T^*X. The set of zeros of this sheaf of ideals is a closed analytic (or quasi-projective) subvariety of T^*X called the **characteristic variety**

$$\mathrm{Ch}(\mathcal{M}) \tag{11.6}$$

of $\mathcal{M}$. Since $\sqrt{\mathcal{I}}$ is generated by elements which are homogeneous polynomials in the variables $\xi_1, \ldots, \xi_n$ of the fibre directions of T^*X it follows that $\mathrm{Ch}(\mathcal{M})$ is a *conical* subvariety of T^*X, i.e. it is invariant under scalar multiplication in the fibres of T^*X.

11.5   Holonomic differential systems

It is easy to prove the following lemma.

Lemma 11.5.1. *If $P \in \mathcal{D}^{(m)}$ and $Q \in \mathcal{D}^{(\ell)}$ then the commutator $[P, Q]$ of P and Q is an element of $\mathcal{D}^{(\ell+m-1)}$ and*

$$\sigma_{\ell+m-1}([P,Q]) = \sum_{1 \leq i \leq n} \frac{\partial f}{\partial \xi_i}\frac{\partial g}{\partial z_i} - \frac{\partial f}{\partial z_i}\frac{\partial g}{\partial \xi_i}$$

where $f = \sigma_m(P)$ and $g = \sigma_\ell(Q)$.

Definition 11.5.2. If f and g lie in $\mathcal{O}[\xi_1, \ldots, \xi_n]$ then the **Poisson bracket** $\{f, g\} \in \mathcal{O}[\xi_1, \ldots, \xi_n]$ of f and g is defined by

$$\{f, g\} = \sum_{1 \leq i \leq n} \frac{\partial f}{\partial \xi_i}\frac{\partial g}{\partial z_i} - \frac{\partial f}{\partial z_i}\frac{\partial g}{\partial \xi_i}.$$

Theorem 11.5.3. *(cf. Pham [143, §2.9], Saito, Kawai and Kashiwara [152]). If $\mathcal{M}$ is a differential system on $\mathbb{C}^n$ and $\sqrt{\mathcal{I}}$ is defined as in Section 11.4 then $\sqrt{\mathcal{I}}$ is involutive, in the sense that if $f \in \sqrt{\mathcal{I}}$ and $g \in \sqrt{\mathcal{I}}$ then $\{f, g\} \in \sqrt{\mathcal{I}}$.*

This theorem can be globalised in the obvious way.

If X is a complex manifold then T^*X has a holomorphic symplectic form ω defined in local coordinates by

$$\omega = \sum_{1 \leq i \leq n} d\xi_i \wedge dz_i.$$

That is, ω is a holomorphic section of $\Lambda^2 T^*(T^*X)$ such that

$$d\omega = 0$$

and when elements of the fibres of $\Lambda^2 T^*(T^*X)$ are identified with skew-symmetric bilinear forms on the tangent spaces to T^*X the skew-symmetric bilinear form defined by ω on each tangent space is non-degenerate.

Theorem 11.5.3 has an infinitesimal counterpart in terms of this symplectic form ω.

Theorem 11.5.4. *Let η be a non-singular point of the characteristic variety $\mathrm{Ch}(\mathcal{M})$ of a differential system $\mathcal{M}$ on X. Then*

$$(T_\eta \mathrm{Ch}(\mathcal{M}))^\perp \subseteq T_\eta \mathrm{Ch}(\mathcal{M}) \tag{11.7}$$

where $\perp$ denotes the orthogonal complement in $T_\eta(T^*X)$ with respect to the holomorphic symplectic form ω.

It follows immediately from Theorem 11.5.4 that if the characteristic variety $\mathrm{Ch}(\mathcal{M})$ is non-empty then its dimension satisfies

$$\dim \mathrm{Ch}(\mathcal{M}) \geq \dim T^* X - \dim \mathrm{Ch}(\mathcal{M}),$$

i.e.

$$\dim \mathrm{Ch}(\mathcal{M}) \geq \dim X. \tag{11.8}$$

Definition 11.5.5. A differential system $\mathcal{M}$ on X is called **holonomic** if $\mathcal{M} = 0$ or

$$\dim \mathrm{Ch}(\mathcal{M}) = n$$

where n is the dimension of X. Equivalently for every non-singular point η of $\mathrm{Ch}(\mathcal{M})$ we have

$$(T_\eta \mathrm{Ch}(\mathcal{M}))^\perp = T_\eta \mathrm{Ch}(\mathcal{M}).$$

Such a subvariety of $T^* X$ is called a **Lagrangian** subvariety of $T^* X$.

Holonomic differential systems used to be called 'maximally over-determined'; 'over-determined' because the number of independent equations is greater than or equal to the number of unknowns (otherwise $\sqrt{\mathcal{I}} = 0$ so the system is not holonomic) and 'maximally' because $\sqrt{\mathcal{I}}$ is as large as possible (equivalently the characteristic variety as is small as possible).

11.6 Examples of characteristic varieties

Consider the differential system $\mathcal{M}_1$ on $\mathbb{C}$ defined by one generator u and one equation

$$(zD - \alpha)^q u = 0 \tag{11.9}$$

where $\alpha \in \mathbb{C}$, $q \in \mathbb{N}$, z is the coordinate on $\mathbb{C}$ and $D = \frac{\partial}{\partial z}$. The filtration

$$\mathcal{M}_1^{(0)} \subseteq \mathcal{M}_1^{(1)} \subseteq \dots$$

of $\mathcal{M}_1$ defined by this choice of generator u is given by

$$\begin{aligned}
\mathcal{M}_1^{(m)} &= \mathcal{D}_1^{(m)} u \\
&= \{a_m(z) D^m u + \dots + a_0(z)u \mid a_j(z) \in \mathcal{O}, \ 0 \leq j \leq m\}.
\end{aligned}$$

Hence $\mathcal{M}_1^{(m)}/\mathcal{M}_1^{(m-1)}$ is generated as an $\mathcal{O}$-module by the image of $D^m u$ in $\mathcal{M}_1^{(m)}/\mathcal{M}_1^{(m-1)}$. The Gr $\mathcal{D}$-module structure on

$$\mathrm{Gr}\mathcal{M}_1 = \bigoplus \mathcal{M}_1^{(m)}/\mathcal{M}_1^{(m-1)}$$

is such that if $p \in \mathcal{D}_1^{(m)}/\mathcal{D}_1^{(m-1)} \subseteq \mathrm{Gr}\mathcal{D}$ is represented by $P \in \mathcal{D}^{(m)}$ and if $x \in \mathcal{M}_1^{(\ell)}/\mathcal{M}_1^{(\ell-1)} \subseteq \mathrm{Gr}\mathcal{M}$ is represented by X then px is the image of PX in

$$\mathcal{M}_1^{(\ell+m)}/\mathcal{M}_1^{(\ell+m-1)} \subseteq \mathrm{Gr}\mathcal{M}.$$

Thus when the symbol is used to identify $\mathrm{Gr}\mathcal{D}$ with $\mathcal{O}[\xi]$ the image of $D^m u$ in $\mathcal{M}_1^{(m)}/\mathcal{M}_1^{(m-1)}$ is $\xi^m u$. Hence $\mathrm{Gr}\mathcal{M}$ is generated as an $\mathcal{O}[\xi]$-module by u. Moreover if

$$a_m(z)\xi^m + \ldots + a_0(z) \in \mathcal{O}[\xi]$$

then

$$a_j(z)\xi^j u \in \mathcal{M}_1^{(j)}/\mathcal{M}_1^{(j-1)} \quad \text{for} \quad 0 \le j \le m$$

so

$$(a^m(z)\xi^m + \ldots + a_0(z))u = 0$$

if and only if

$$a_j(z)\xi^j u = 0$$

for $0 \le j \le m$, and this happens if and only if

$$a_j(z)D^j u \in \mathcal{M}_1^{(j-1)}$$

for $0 \le j \le m$. But $a_j(z)D^j u \in \mathcal{M}_1^{(j-1)}$ if and only if there exist

$$b_0(z), \ldots, b_{j-1}(z) \in \mathcal{O}$$

such that

$$a_j(z)D^j u = b_{j-1}(z)D^{j-1}u + \ldots + b_0(z)u,$$

i.e. if and only if $a_j(z)\xi^j$ is the symbol for some $P \in \mathcal{D}$ such that $Pu = 0$. In our case $Pu = 0$ if and only if

$$P \in \mathcal{D}(zD - \alpha)^q$$

so $a_j(z)\xi^j u = 0$ if and only if $(z\xi)^q$ divides $a_j(z)\xi^j$. Thus $\mathrm{Gr}\mathcal{M}_1$ is generated as $\mathcal{O}[\xi]$-module by one generator u with relaxation

$$(z\xi)^q u = 0.$$

Thus the annihilator $\mathcal{I}$ of $\mathrm{Gr}\mathcal{M}_1$ in $\mathcal{O}[\xi]$ is the ideal generated by $(z\xi)^q$ and $\sqrt{\mathcal{I}}$ is generated by $z\xi$. Thus

$$
\begin{aligned}
\mathrm{Ch}(\mathcal{M}) &= \{(z,\xi) \in T^*\mathbb{C} \,|\, z\xi = 0\} \\
&= \{(z,\xi) \in T^*\mathbb{C} \,|\, z = 0 \text{ or } \xi = 0\}.
\end{aligned}
$$

We can choose a different set of generators and relations for $\mathcal{M}_1$ as follows. Let

$$u_j = (zD - \alpha)^{j-1}u \text{ for } 1 \le j \le q.$$

Then $u_1 = u$ so $u_1, \ldots, u_q$ generate $\mathcal{M}_1$ with relations

$$
\begin{aligned}
zDu_j &= u_{j+1} + \alpha u_j, \quad 1 \le j < m, \\
zDu_m &= \alpha u_m.
\end{aligned}
$$

With this set of generators $\mathrm{Gr}\mathcal{M}_1$ becomes the $\mathcal{O}[\xi]$-module generated by $u_1, \ldots, u_q$ with relations

$$z\xi u_j = 0, \qquad 1 \leq j \leq m.$$

Thus in this case both $\mathcal{I}$ and $\sqrt{\mathcal{I}}$ are the ideal generated by $z\xi$.

$\mathcal{M}_1$ extends to a different system $\mathcal{M}$ on $\mathbb{CP}^1 = \mathbb{C} \cup \{\infty\}$ as follows. Let w be the local coordinate on $\mathbb{CP}^1 - \{0\}$ given by

$$w = z^{-1}$$

for $z \in \mathbb{C} - \{0\}$ and such that w takes the value 0 at ∞. Then on $\mathbb{C} - \{0\}$

$$zD = z\frac{d}{dz} = z\frac{dw}{dz}\frac{d}{dw} = -\frac{1}{z}\frac{d}{dw} = -wD_w$$

where $D_w = \frac{d}{dw}$. Thus if $\mathcal{M}_2$ is the differential system on $\mathbb{CP}^1 - \{0\}$ defined by one generator u and one relation

$$(wD_w + \alpha)^q u = 0$$

then there is an obvious isomorphism between the restrictions of $\mathcal{M}_1$ and $\mathcal{M}_2$ to $\mathbb{C} - \{0\}$. Hence there is a $\mathcal{D}_X$-module $\mathcal{M}$ on $X = \mathbb{CP}^1$ such that

$$\mathcal{M}|_{\mathbb{C}} \cong \mathcal{M}_1, \qquad \mathcal{M}|_{\mathbb{CP}^1 - \{0\}} \cong \mathcal{M}_2.$$

The characteristic variety $\mathrm{Ch}(\mathcal{M})$ is the subvariety of $T^*\mathbb{CP}^1$ which is the union of the zero section and the fibres over 0 and ∞. Since

$$\dim \mathrm{Ch}(\mathcal{M}) = 1 = \dim \mathbb{CP}^1,$$

$\mathcal{M}$ is holonomic.

Remark 11.6.1. Of course when X is 1-dimensional a coherent $\mathcal{D}_X$-module $\mathcal{M}$ is holonomic if and only if its characteristic variety $\mathrm{Ch}(\mathcal{M})$ is not equal to T^*X, or equivalently the sheaf of ideals $\sqrt{\mathcal{I}}$ is non-zero. In particular a non-zero differential system on X defined locally by one generator and one non-zero equation is always holonomic.

The differential system $\mathcal{M}$ on $\mathbb{C}^2$ defined by one generator u and one equation

$$D_1^2 u + D_2^2 u + a(z_1, z_2)u = 0 \tag{11.10}$$

has characteristic variety $\mathrm{Ch}(\mathcal{M})$ defined by

$$\xi_1^2 + \xi_2^2 = 0,$$

in $T^*\mathbb{C}^2$. Thus $\dim \mathrm{Ch}(\mathcal{M}) = 3 > \dim \mathbb{C}^2$ so $\mathcal{M}$ is not holonomic.

11.7   Left and right $\mathcal{D}_X$-modules

The sheaf of holomorphic (respectively regular) vector fields on a non-singular complex variety X (i.e. holomorphic or regular sections of TX) can be regarded as an $\mathcal{O}_X$-submodule of the sheaf of rings $\mathcal{D}_X$. In local coordinates a vector field

$$a_1(z)\frac{\partial}{\partial z_1} + \ldots + a_n(z)\frac{\partial}{\partial z_n},$$

where the $a_i(z)$ are holomorphic or regular functions of $z = (z_1, \ldots, z_n)$, is identified with the differential operator

$$a_1(z)D_1 + \ldots + a_n(z)D_n$$

given by differentiation along the vector field. As a sheaf of rings $\mathcal{D}_X$ is generated by these vector fields together with $\mathcal{O}_X$. Thus a $\mathcal{D}_X$-module structure on an $\mathcal{O}_X$-module $\mathcal{M}$ is determined by the action on $\mathcal{M}$ of these vector fields.

We have been working with left $\mathcal{D}_X$-modules but we can go freely between left $\mathcal{D}_X$-modules and right $\mathcal{D}_X$-modules. (For more details see e.g. Pham [143, §2.13]). If v is a holomorphic vector field on an open subset U of X and if ω is a holomorphic n-form on U (i.e. a holomorphic section of $\Lambda^n T^* X$ over U) then we can contract v and ω using the dual paring between TX and $T^* X$ to get a holomorphic $(n-1)$-form $\iota_v\omega$ on U. Clearly $\iota_v\omega$ is regular if both v and ω are regular. If $n = \dim X$ then the Lie derivative of ω along v is the n-form

$$\mathrm{Lie}_v(\omega) = d(\iota_v\omega).$$

Given a left $\mathcal{D}_X$-module $\mathcal{M}$ we can put a right $\mathcal{D}_X$-module structure on the tensor product

$$\Omega(\mathcal{M}) = \Omega_X^n \otimes_{\mathcal{O}_X} \mathcal{M}$$

where Ω_X^n is the sheaf of holomorphic (or regular) sections of $\Lambda^n T^* X$, as follows. If U is an open subset of X and if $\omega \in \Omega_X^n(U)$, $u \in \mathcal{M}(U)$, $f \in \mathcal{O}_X(U) \subseteq \mathcal{D}_X(U)$ and v is a holomorphic (or regular) vector field on U define

$$(\omega \otimes u)f \;=\; \omega \otimes fu, \qquad\qquad (11.11)$$
$$(\omega \otimes u)z \;=\; -\mathrm{Lie}_v(\omega) \otimes u - \omega \otimes vu. \qquad (11.12)$$

This defines a right $\mathcal{D}_X$-module structure on $\Omega(\mathcal{M})$.

The motivation for this definition is the usual process of identifying functions with distributions by multiplying by a fixed differential form of top degree. It is not hard to check (see e.g. Bernstein [15, Lect. 1 §4], Borel [20, VI §3], Pham [143, §2.13]) that Ω is a functor which induces an equivalence of the category of left $\mathcal{D}_X$-modules with the category of right $\mathcal{D}_X$-modules.

11.8   Restriction of $\mathcal{D}_X$-modules

Let Y be a non-singular subvariety of a non-singular variety X, let $i : Y \to X$ be the inclusion and let $\mathcal{M}$ be a $\mathcal{D}_X$-module. One would like to be able to restrict the $\mathcal{D}_X$-module $\mathcal{M}$ to give a $\mathcal{D}_Y$-module in some sensible way. If Y is an open subset of X then this is easy (because then open subsets of Y are open subsets of X) so we may as well assume Y is a closed subvariety of X.

It is not hard to restrict $\mathcal{M}$ as a *sheaf* on X to a sheaf $\mathcal{M}|_Y$ on Y. If U is an open subset of Y we set

$$\mathcal{M}|_Y(U) = \operatorname{colim} \mathcal{M}(V)$$

where the limit is over all open subsets V of X containing U. Then if $y \in Y$ the stalk of $\mathcal{M}|_Y$ at y is the same as the stalk of $\mathcal{M}$ at y.

The sheaf $\mathcal{M}|_Y$ is an $\mathcal{O}_X|_Y$-module, and so is the sheaf $\mathcal{O}_Y$. The tensor product

$$i^\circ \mathcal{M} = \mathcal{O}_Y \otimes_{\mathcal{O}_X|_Y} \mathcal{M}|_Y \tag{11.13}$$

has a natural $\mathcal{O}_Y$-module structure given by $f(g \otimes u) = fg \otimes u$. In order to make $i^\circ \mathcal{M}$ a $\mathcal{D}_Y$-module it is necessary to define the action of holomorphic vector fields (cf. Section 11.7).

Suppose that U is an open subset of Y and $u \in \mathcal{M}|_Y(U)$. Then there is an open subset of V of X such that $U = V \cap Y$ and

$$u = \tilde{u}|_Y$$

for some $\tilde{u} \in \mathcal{M}(V)$. By choosing U and V small enough we can assume that there are coordinates $z_1, \dots, z_n$ on V and $y_1, \dots, y_m$ on U. If

$$v = \sum_{i \leq i \leq m} a_i(y) \frac{\partial}{\partial y_i}$$

is a holomorphic vector field on U and $f \in \mathcal{O}_Y(U)$ is a holomorphic function on U then $v(f)$ is the holomorphic function on U given by

$$v(f) = \sum_{1 \leq i \leq m} a_i(y) \frac{\partial f}{\partial y_i}.$$

Now define the action of v on the element of $f \otimes u$ of $i^\circ \mathcal{M}(U)$ by

$$v(f \otimes u) = v(f) \otimes u + \sum_{1 \leq i \leq n} f v(z_i) \otimes \left. \frac{\partial \tilde{u}}{\partial z_i} \right|_y. \tag{11.14}$$

It can be checked (see e.g. Borel [20, VI, 4.1]) that this action is independent of the choice of coordinates and defines a $\mathcal{D}_Y$-module structure on $i^\circ \mathcal{M}$. In a similar way given any holomorphic map $\pi \colon Y \to X$ and $\mathcal{D}_X$-module $\mathcal{M}$ we can define a $\mathcal{D}_Y$-module $\pi^\circ \mathcal{M}$.

Example 11.8.1. Suppose $\mathcal{M} = \mathcal{D}_X$. Locally we can choose coordinates $z_1, \ldots, z_n$ such that Y is defined by

$$z_1 = z_2 = \ldots = z_d = 0.$$

Then $i^\circ \mathcal{M}$ is the locally free $\mathcal{D}_Y$-module with local basis the set of all monomials in $D_1, \ldots, D_d$ where

$$D_i = \frac{\partial}{\partial z_i}.$$

Example 11.8.2. Let $X = \mathbb{C}^2$ and let

$$Y = \{(z_1, z_2) \in \mathbb{C}^2 | z_2 = z_1^2\}.$$

We can identify Y with $\mathbb{C}$ via the isomorphism

$$z \in \mathbb{C} \mapsto (z, z^2) \in Y.$$

Let $\mathcal{M}$ be the $\mathcal{D}_X$-module with one generator u and two relations

$$\begin{aligned}(D_1^2 + D_2^2)u &= 0, \\ z_1 u &= 0.\end{aligned}$$

We can change coordinates on $\mathbb{C}^2$ from (z_1, z_2) to (z, w) where $z = z_1$ and $w = z_2 - z_1^2$. Then $z_1 = z$ and $z_2 = w + z^2$ so

$$\frac{\partial}{\partial z} = D_1 + 2z D_2, \qquad \frac{\partial}{\partial w} = D_2.$$

Hence

$$D_2^j(D_1^2 + D_2^2) = D_2^j(\frac{\partial}{\partial z} + 2z D_2)^2 + D_2^{j+2}$$

$$= \left(\frac{\partial}{\partial z}\right)^2 D_2^j - \left(4z\left(\frac{\partial}{\partial z}\right) + 2\right) D_2^{j+1} + (1 + 4z^2) D_2^{j+2}$$

and

$$D_2^j z_1 = z_1 D_2^j.$$

Thus $i^\circ(\mathcal{M})$ is the quotient of the free $\mathcal{D}_Y$-module with basis

$$\{D_2^j u | j \geq 0\}$$

by the submodule generated by

$$\{D^2 D_2^j u - (2 + 4z D) D_2^{j+1} u + (1 + 4z^2) D_2^{j+2} u | j \geq 0\}$$

and

$$\{z D_2^j u | j \geq 0\}$$

where z is the standard coordinate on $Y \cong \mathbb{C}$ and

$$D = \frac{d}{dz}.$$

Equivalently $i^\circ(\mathcal{M})$ is generated as a $\mathcal{D}_Y$-module by u and $v = D_2 u$ subject to the relations

$$zu = 0 = zv, \qquad Du = 0 = Dv.$$

Remark 11.8.3. Suppose that F is a left exact covariant functor from the category of quasi-coherent $\mathcal{D}_X$-modules to the category of quasi-coherent $\mathcal{D}_Y$-modules. (Quasi-coherence is a technical condition which is weaker than coherence; for the definition see Bernstein [15, §1.1], Borel [20, VI ,1.4], Hartshorne [79, II §5]). Recall the definition of right derived functor given in Section 3.8: given a $\mathcal{D}_X$-module $\mathcal{M}$ we choose an injective resolution

$$0 \longrightarrow \mathcal{M} \longrightarrow \mathcal{I}_0 \overset{d_0}{\longrightarrow} \mathcal{I}_1 \overset{d_1}{\longrightarrow} \mathcal{I}_2 \longrightarrow \cdots$$

of $\mathcal{M}$ (as a quasi-coherent $\mathcal{D}_X$-module) and define $RF(\mathcal{M})$ to be the complex

$$0 \longrightarrow F(\mathcal{I}_0) \overset{F(d_0)}{\longrightarrow} F(\mathcal{I}_1) \overset{F(d_1)}{\longrightarrow} F(\mathcal{I}_2) \longrightarrow \cdots .$$

If F is left exact then this complex is independent of the choice of injective resolution up to quasi-isomorphism. Similarly if F is right exact then we can define the left derived functor LF of F by using a projective resolution and reversing all arrows in the definitions. $LF(\mathcal{M})$ is a complex of $\mathcal{D}_Y$-modules defined up to quasi-isomorphism.

Sometimes it is convenient to think of the restriction of a $\mathcal{D}_X$-module $\mathcal{M}$ to Y as a complex of $\mathcal{D}_Y$-modules rather than a single $\mathcal{D}_Y$-module. We can regard i° as a right exact covariant functor from the category of quasi-coherent $\mathcal{D}_X$-modules to the category of quasi-coherent $\mathcal{D}_Y$-modules, and thus we can consider its left derived functor Li°. It turns out to be convenient to make a dimension shift and so we define

$$i^!(\mathcal{M}) = Li^\circ(\mathcal{M})[d]$$

where

$$d = \dim X - \dim Y$$

(Borel [20, VI §4.2], Bernstein [15, §1.8]). It is often useful to regard either the complex $i^!\mathcal{M}$ or the $\mathcal{D}_Y$-module $\mathcal{H}^0(i^!\mathcal{M})$ as the restriction of $\mathcal{M}$ to Y. From our point of view the main reason for this is the following theorem.

Theorem 11.8.4. *(Kashiwara). Let $i : Y \to X$ be a closed embedding of non-singular varieties. Then the functor*

$$\mathcal{M} \to \mathcal{H}^0(i^!\mathcal{M})$$

is an equivalence between the category of holonomic $\mathcal{D}_X$-modules with support in Y and the category of holonomic $\mathcal{D}_Y$-modules.

For the proof of this theorem see Bernstein [15, §1.10, §3.1] or Borel [20, VI §7.11].

11.9   Regular singularities

Definition 11.9.1. A $\mathcal{D}_X$-module $\mathcal{M}$ on a non-singular variety X is called a **connection** if it is a coherent locally free $\mathcal{O}_X$-module for the $\mathcal{O}_X$-structure coming from the embedding $\mathcal{O}_X \to \mathcal{D}_X$.

Remark 11.9.2. In fact a $\mathcal{D}_X$-module which is coherent as an $\mathcal{O}_X$-module is locally free as an $\mathcal{O}_X$-module (Bernstein [15, §2.1(a)], Borel [20, IV §1.1]).

Remark 11.9.3. Let $\mathcal{M}$ be a coherent locally free $\mathcal{O}_X$-module. Then locally $\mathcal{M}$ is freely generated as an $\mathcal{O}_X$-module by finitely many sections $u_1, \ldots, u_m$, say. This means that $\mathcal{M}$ can be identified with the sheaf of holomorphic (or regular) sections of a complex vector bundle V of rank m over X. If moreover $\mathcal{M}$ is a $\mathcal{D}_X$-module then there exist local sections Γ_{ij}^k of $\mathcal{O}_X$ such that

$$D_i u_j = \sum_{1 \le k \le m} \Gamma_{ij}^k u_k.$$

The local sections Γ_{ij}^k define a flat connection on V in the sense of differential geometry. Flatness corresponds to the commutativity conditions

$$[D_i, D_j] = 0.$$

Remark 11.9.4. Let $\mathcal{M}$ be a connection on X generated locally as an $\mathcal{O}_X$-module by a basis of sections $\{u_1, \ldots, u_m\}$. Then $\mathrm{Gr}\mathcal{M}$ is generated locally as an $\mathcal{O}_X[\xi_1, \ldots, \xi_n]$-module by $u_1, \ldots, u_m$ with relations

$$\xi_i u_j = 0, \qquad 1 \le i \le n, \qquad 1 \le j \le m.$$

Thus $\sqrt{\mathcal{I}}$ is locally the sheaf of ideals generated by $\xi_1, \ldots, \xi_n$ in $\mathcal{O}_X[\xi_1, \ldots, \xi_n]$. Hence the characteristic variety $\mathrm{Ch}(\mathcal{M})$ of $\mathcal{M}$ is the zero section in T^*X. In particular any connection is a holonomic $\mathcal{D}_X$-module.

Example 11.9.5. Consider the differential system $\mathcal{M}$ on $\mathbb{C}$ defined by one generator u and one equation

$$(zD - \alpha)^m u = 0$$

where $\alpha \in \mathbb{C}$ and $m \in \mathbb{N}$. Then $\mathcal{M}\big|_{\mathbb{C} - \{0\}}$ is generated by the global sections

$$u_1 = u, \qquad u_j = (zD - \alpha)^{j-1} u, \qquad 1 < j \le m,$$

with relations

$$Du_j = \frac{1}{z} u_{j+1} + \frac{\alpha}{z} u_j, \qquad 1 \le j < m \tag{11.15}$$

$$Du_m = \frac{\alpha}{z} u_m. \tag{11.16}$$

If α is not an element of $\mathbb{N}$ the space of solutions of $\mathcal{M}$ in $\mathcal{O}_X$ over any simply connected subset of $\mathbb{C} - \{0\}$ is spanned by the solutions

$$u = z^\alpha (\log z)^j, \qquad 0 \le j \le m - 1.$$

$\mathcal{M}\big|_{\mathbb{C}-\{0\}}$ is a connection. But $\mathcal{M}$ has no sections over any neighbourhood of 0 in $\mathbb{C}$ so $\mathcal{M}$ itself is not a connection. It is the presence of z^{-1} factors in the relations (11.15) and (11.16) which prevents $\mathcal{M}$ from being a connection over $\mathbb{C}$.

If X is quasi-projective and non-singular of dimension one (i.e. X is a non-singular curve) then we can choose an embedding of X in a non-singular projective curve X^+ such that $X^+ - X$ is a finite set of points. Fix $s \in X^+$ and choose a local coordinate z on a neighbourhood U of s in X^+ such that z vanishes at s. Let $D = \frac{d}{dz}$ be the corresponding differential operator on U.

Definition 11.9.6. Let $\mathcal{M}$ be a holonomic $\mathcal{D}_X$-module. Then $\mathcal{M}$ has a **regular singularity** at s if U and z can be chosen such that

$$\mathcal{M}\big|_{U-\{s\}}$$

is a connection on $U - \{s\}$ and is generated as a $\mathcal{D}_{U-\{s\}}$-module by a finitely generated $\mathcal{O}_U$-module which is invariant under the action of zD. That is, on $U - \{s\}$ the module $\mathcal{M}$ is defined by a system of equations in variables $u_1, \ldots, u_p$ such that for all i we can write

$$zDu_i = \sum_{1 \leq j \leq p} a_{ij} u_j$$

where $a_{ij} \in \mathcal{O}_X(U)$. Equivalently

$$Du_i = \sum_{1 \leq j \leq p} \frac{a_{ij}}{z} u_j$$

where the a_{ij} extend to holomorphic functions of z on U.

In fact a system $\mathcal{M}$ on $\mathbb{C}$ has a regular singularity at 0 if and only if near 0 it is isomorphic to a finite direct sum of $\mathcal{D}$-modules of the form

$$\mathcal{D}/\mathcal{D}(zD - \alpha)^m$$

for $m \in \mathbb{N}$ and $\alpha \in \mathbb{C} - \mathbb{N}$ (see e.g. Pham [143, §2.11.6]).

Definition 11.9.7. The holonomic $\mathcal{D}_X$-module $\mathcal{M}$ has **regular singularities** if it has a regular singularity at each $s \in X^+$. It is not hard to check that this definition is independent of the choice of X^+.

Definition 11.9.8. A complex of $\mathcal{D}_X$-modules is holonomic (and has regular singularities) if all its cohomology sheaves are holonomic (and have regular singularities).

Finally if $\mathcal{M}$ is a holonomic $\mathcal{D}_X$-module on a non-singular variety X of any dimension we make the following definition.

Definition 11.9.9. $\mathcal{M}$ has regular singularities if, for any non-singular curve C in X whose inclusion map $i : C \hookrightarrow X$ is a closed embedding, the restriction $i^! \mathcal{M}$ of $\mathcal{M}$ has regular singularities.

11.10   The Riemann–Hilbert correspondence

Definition 11.10.1. Let A be a non-singular subvariety of a non-singular variety X. The **conormal bundle** to A in X is

$$T_A^* X = \{y \in T^* X \,|\, \pi(y) \in A, \quad y \in (T_{\pi(y)} A)^\circ\}$$

where $\pi : T^* X \to X$ is the projection and $(T_{\pi(y)} A)^\circ$ is the annihilator of the tangent space $T_{\pi(y)} A$ to A at $\pi(y)$ in the dual $T^*_{\pi(y)} X$ of $T_{\pi(y)} X$.

If A is a singular subvariety of X with

$$\widetilde{A} = \{\text{non-singular points of } A\}$$

then we define $T_A^* X$ to be the closure in $T^* X |_A$ of $T_{\widetilde{A}}^* X$. Note that $T_X^* X$ is the zero section of $T^* X$.

Proposition 11.10.2. *(see e.g. Pham [143, §2.10.1]). Let $V \subseteq T^* X$ be an irreducible Lagrangian conical closed subvariety of $T^* X$. Then the image $\pi(V)$ of V under $\pi : T^* X \to X$ is an irreducible subvariety of X and*

$$V = T_{\pi(V)}^* X$$

is the conormal bundle to $\pi(V)$ in X.

Recall from Sections 11.4 and 11.5 that if $\mathcal{M}$ is a holonomic $\mathcal{D}_X$-module then the characteristic variety $\mathrm{Ch}(\mathcal{M})$ of $\mathcal{M}$ is a closed conical Lagrangian subvariety of $T^* X$.

Corollary 11.10.3. *If $\mathcal{M}$ is a holonomic $\mathcal{D}_X$-module then every irreducible component V of $\mathrm{Ch}(\mathcal{M})$ is of the form*

$$V = T_S^* X$$

where S is an irreducible subvariety of X.

Lemma 11.10.4. *(Pham [143, §2.10.3]). Let $\mathcal{M}$ be a holonomic $\mathcal{D}_X$-module. If $V_1, \ldots, V_p$ are the irreducible components of $\mathrm{Ch}(\mathcal{M})$ and if $V_i = T_{S_i}^* X$ let*

$$S = \bigcup_{S_i \neq X} S_i.$$

Then the restriction of $\mathcal{M}$ to $X - S$ is a connection (possibly zero).

Let $\mathcal{M}$ be a connection on a quasi-projective variety X. The **sheaf of horizontal sections** of $\mathcal{M}$ is the sheaf $\mathcal{F}$ on X such that if $(z_1, \ldots, z_n)$ are local coordinates on an open subset of V of X then

$$\mathcal{F}(V) = \{u \in \mathcal{M}(V) \,|\, D_i u = 0, \quad 1 \leq i \leq n\}.$$

It is easy to check that this is a good definition, independent of the choice of local coordinates. Now suppose that $\mathcal{M}$ is freely generated as an $\mathcal{O}_X$-module over V by sections $u_1, \ldots, u_m$ of $\mathcal{M}(V)$ with

$$D_i u_j = \sum_{1 \leq k \leq m} \Gamma_{ij}^k u_k.$$

Then a general element u of $\mathcal{F}(V)$ can be written in the form

$$u = f_1 u_1 + \ldots + f_m u_m$$

where $f_1, \ldots, f_m \in \mathcal{O}_X(V)$ are functions on V satisfying

$$\frac{\partial f_j}{\partial z_i} + \sum_{1 \leq k \leq m} \Gamma_{ik}^j f_k = 0$$

for $1 \leq i \leq n$ and $1 \leq j \leq m$. The theory of existence and uniqueness of local solutions of differential equations implies that if V is simply connected then the restriction to V of $\mathcal{F}$ is isomorphic to the constant sheaf $\mathbb{C}_V^m$, or equivalently the restriction map

$$\mathcal{F}(V) \to \mathcal{F}_X$$

is an isomorphism for all $x \in V$. This means that $\mathcal{F}$ defines a local system $\mathcal{L}$ on X with $\mathcal{L}_X = \mathcal{F}_X$ (cf. Section 4.9).

We can now give a classification of irreducible holonomic $\mathcal{D}_X$-modules which is itself sometimes called the Riemann–Hilbert correspondence (cf. Borel [20, IV], Deligne [52], Bernstein [15, §3.14 and §4.1]).

Theorem 11.10.5.

(i) *Let Y be a closed irreducible subvariety of a non-singular variety X, and let $\mathcal{L}$ be an irreducible local system on a dense open non-singular subvariety U of Y. Then there is a unique irreducible holonomic $\mathcal{D}_X$-module with regular singularities, denoted $\mathcal{M}(Y, \mathcal{L})$, whose support is contained in Y and whose restriction to U is a connection such that the local system defined by its sheaf of horizontal section is $\mathcal{L}$.*

(ii) *Any irreducible holonomic $\mathcal{D}_X$-module with regular singularities is isomorphic to $\mathcal{M}(Y, \mathcal{L})$ for some Y and $\mathcal{L}$ as in (i).*

(iii) *$\mathcal{M}(Y', \mathcal{L}')$ is isomorphic to $\mathcal{M}(Y, \mathcal{L})$ if and only if $Y = Y'$ and the restrictions of $\mathcal{L}$ and $\mathcal{L}'$ to some non-empty open subset of Y, on which they are both defined, are isomorphic.*

In order to relate this form of the Riemann–Hilbert correspondence to one which involves intersection cohomology, we need the following theorem (cf. Bernstein [15, §5.9]).

Theorem 11.10.6. *Let Y be a closed irreducible subvariety of a non-singular n-dimensional variety X and let $\mathcal{L}$ be a local system on a dense open non-singular subvariety of Y. Then the de Rham complex*

$$0 \to \mathcal{M}(Y, \mathcal{L}) \to \Omega^1_X \otimes_{\mathcal{O}_X} \mathcal{M}(Y, \mathcal{L}) \to \ldots \to \Omega^j_X \otimes_{\mathcal{O}_X} \mathcal{M}(Y, \mathcal{L}) \to \ldots$$

of $\mathcal{M}(Y, \mathcal{L})$ has support in Y and its restriction to Y is isomorphic in the derived category $D^b(X)$ to the intersection sheaf complex $\mathcal{IC}^{\bullet}_{(Y,\mathcal{L})}$ with a shift in degree.

The idea of the proof of Theorem 11.10.6 is to check that, after a shift in degree, $\mathrm{DR}(\mathcal{M}(Y, \mathcal{L}))$ satisfies the conditions which uniquely characterise $\mathcal{IC}^{\bullet}_{(Y,\mathcal{L})}$ (cf. Section 7.3).

Using Theorems 11.10.5 and 11.10.6 one can obtain the Riemann–Hilbert correspondence in the following form, first proved by Kashiwara [91] and Mebkhout [131] in the holomorphic case and Beilinson and Bernstein in the algebraic case (see Bernstein [15] and Borel [20, VIII]).

Theorem 11.10.7. *The de Rham functor DR induces a one-to-one correspondence between isomorphism classes of irreducible holonomic $\mathcal{D}_X$-modules with regular singularities and isomorphism classes (in $D^b(X)$) of intersection cohomology complexes of irreducible closed subvarieties of X with coefficients in irreducible local systems.*

Remark 11.10.8. In fact the de Rham functor gives an equivalence of categories between the category of holonomic $\mathcal{D}$-modules with regular singularities on X and the derived category of perverse sheaves of X (Borel [20, VIII 14.4], Bernstein [15, §5.9]). (The derived category of perverse sheaves is obtained from the category of perverse sheaves by formally inverting all quasi-isomorphisms, so that they become isomorphisms.) By considering only the irreducible objects in each category this gives the Riemann–Hilbert correspondence in the form above.

11.11 Further reading

Deligne's paper [52] was pivotal in the formulation of the modern version of the Riemann–Hilbert correspondence. Mebkhout [131] and Kashiwara [91] give proofs of the Riemann–Hilbert correspondence for analytic $\mathcal{D}$-modules. Borel [20] contains a detailed treatment of the Riemann–Hilbert correspondence for algebraic $\mathcal{D}$-modules (see also Bernstein [15]). A good reference for the analytic version of the theory is Kashiwara [95]; see also Björk [19], Malgrange [127] and Mebkhout [132]. Dimca [56] and Gelfand and Manin [63, Chapter 8] contain more condensed expositions. Tráng and Mebkhout [173] is an introductory survey and Oda [140] an 'elementary' survey in the case when the underlying space is a complex manifold.

The notion of characteristic variety belongs to micro-local analysis, for further details on this topic see e.g. Kashiwara [92], [95] and Kashiwara and Schapira [97]. Kashiwara, Kawai, and Kimua [96] is another approach which avoids the apparatus of derived categories. More recently there have been attempts to prove a micro-local version of the Riemann–Hilbert correspondence, see Andronikof [2, 3], Neto [138] and Waschkies [176].

Chapter 12

The Kazhdan–Lusztig conjecture

In this chapter we shall describe briefly the proof of a conjecture of Kazhdan and Lusztig [104], [105] concerning the representation theory of Lie algebras. The proof (following Bernstein [15]) involves translating the problem first into the language of $\mathcal{D}$-modules, then via the Riemann–Hilbert correspondence into a problem involving intersection cohomology and finally, using ℓ-adic intersection cohomology, into the theory of Hecke algebras.

First it is necessary to review some basic facts about the representation theory of complex Lie groups and Lie algebras. For more details see e.g. Atiyah *et al.* [4], Bourbaki [27], Chevalley [49], Jacobson [85], Kac [88] and Springer [165].

12.1 Verma modules

Let K be a compact Lie group. For simplicity let us assume that K is connected and simply connected. Let $\mathfrak{k}$ be the Lie algebra of K and let

$$\mathfrak{g} = \mathfrak{k} \otimes_{\mathbb{R}} \mathbb{C}$$

be its complexified Lie algebra. The Lie bracket $[\ , \ ]$ on $\mathfrak{g}$ is the unique complex bilinear extension of the Lie bracket on $\mathfrak{k}$.

There is a unique connected, simply connected complex Lie group G whose Lie algebra is $\mathfrak{g}$. We shall assume for simplicity that G is semisimple; that is, that its Lie algebra $\mathfrak{g}$ has no non-zero Abelian ideals. For many reasons mathematicians have long been interested in the complex representations of such complex Lie groups G and their Lie algebras $\mathfrak{g}$.

Let T be a maximal torus of G and let $N_G(T)$ be its normaliser in G. Then

$$W = N_G(T)/T$$

is a finite group called the **Weyl group** of G.

Example 12.1.1. We can take

$$
\begin{aligned}
K &= \mathrm{SU}(n), \\
G &= \mathrm{SL}(n, \mathbb{C}), \\
T &= \{\text{diagonal matrices in } \mathrm{SL}(n, \mathbb{C})\}, \\
W &\cong \Sigma_n
\end{aligned}
$$

where Σ_n denotes the symmetric group.

Let $\mathfrak{h}$ be the Lie algebra of T and let $\mathfrak{h}^*$ be its dual vector space. Then $\alpha \in \mathfrak{h}^* - 0$ is called a **root** of $\mathfrak{g}$ if there exists some non-zero $\xi \in \mathfrak{g}$ such that

$$[h, \xi] = \alpha(h)\xi$$

for all $h \in \mathfrak{h}$. Let $\mathfrak{g}^\alpha$ be the set of all $\xi \in \mathfrak{h}$ such that

$$[h, \xi] = \alpha(h)\xi$$

for all $h \in \mathfrak{h}$. Let Σ be the set of roots of $\mathfrak{g}$. Then

$$\mathfrak{g} = \mathfrak{h} \oplus \Big(\bigoplus_{\alpha \in \Sigma} \mathfrak{g}_\alpha \Big).$$

The Weyl group W acts on $\mathfrak{h}$ and $\mathfrak{h}^*$ and permutes the roots. W is generated by elements which act as reflections in hyperplanes. We can choose a fundamental domain (called a **Weyl chamber**) for the action on W on $\mathfrak{h}^*$ which is a cone in $\mathfrak{h}^*$ bounded by hyperplanes.

Let $\mathfrak{h}^*_+$ be the chosen Weyl chamber (called the **positive Weyl chamber**). Then $\alpha \in \Sigma$ is called a positive root if

$$\alpha(x) > 0$$

for all x in the interior of $\mathfrak{h}^*_+$. Let Σ^+ be the set of positive roots. Then Σ is the disjoint union of Σ^+ and $-\Sigma^+$. Let

$$\mathfrak{N} = \bigoplus_{\alpha \in \Sigma_+} \mathfrak{g}^\alpha.$$

Then $\mathfrak{N}$ is a nilpotent subalgebra of $\mathfrak{g}$. There is a partial order on $\mathfrak{h}^*$ defined by

$$\alpha \geq \beta \Leftrightarrow \alpha(x) \geq \beta(x) \quad \forall x \in \mathfrak{h}_+.$$

Now let M be a $\mathfrak{g}$-module, that is, a complex representation of $\mathfrak{g}$. Then M is a complex vector space (possibly infinite-dimensional) with an action of $\mathfrak{g}$ given by a complex linear map

$$\rho : \mathfrak{g} \to \mathrm{End}(M) = \{\alpha : M \to M \mid \alpha \text{ complex linear}\}$$

which takes the Lie bracket on $\mathfrak{g}$ to the usual Lie bracket

$$[\alpha, \beta] = \alpha\beta - \beta\alpha$$

on $\text{End}(M)$. Assume that M is finitely-generated, i.e. that there exist $m_1, ..., m_k \in M$ such that the only $\mathfrak{g}$-submodule of M containing $m_1, ..., m_k$ is M itself.

If we restrict the representation ρ of $\mathfrak{g}$ to the Abelian subalgebra $\mathfrak{h}$ (or equivalently think of M as an $\mathfrak{h}$-module) then M decomposes as a direct sum

$$M = \bigoplus_{\chi \in h^*} M^\chi$$

where

$$M^\chi = \{m \in M \mid hm = \chi(h)m \; \forall h \in \mathfrak{h}\}.$$

χ is called a **weight** of M if $M^\chi \neq 0$ and a **highest weight** if in addition $\eta \leq \chi$ whenever $M^\eta \neq 0$. If $\alpha \in \Sigma$ and $\xi \in \mathfrak{g}^\alpha$ and $m \in M^\chi$ and $h \in \mathfrak{h}$ then

$$\begin{aligned} h(\xi m) &= [h, \xi]m + \xi(hm) \\ &= \alpha(h)m + \xi(\chi(h)m) \\ &= (\alpha + \chi)(h)\xi m \end{aligned}$$

so

$$\mathfrak{g}^\alpha M^\chi \subseteq M^{\alpha + \chi}.$$

But if $\alpha \in \Sigma^+$ then $\alpha + \chi > \chi$. Thus if χ is a highest weight and $\alpha \in \Sigma^+$ then $M^{\alpha + \chi} = 0$ so

$$\mathfrak{g}^\alpha M^\chi = 0.$$

Hence if χ is a highest weight then

$$\mathfrak{N}M^\chi = 0.$$

M is called a **highest weight $\mathfrak{g}$-module** if it is generated by a single element $m \in M^\chi$ where χ is a highest weight. Any finitely generated $\mathfrak{g}$-module has a filtration

$$M = M_0 \supseteq M_1... \supseteq M_q = 0$$

by $\mathfrak{g}$-module M_j such that the quotient $\mathfrak{g}$-modules M_j/M_{j+1} are all highest weight modules.

Proposition 12.1.2. *For each $\chi \in \mathfrak{h}^*$ there exists a unique (usually infinite-dimensional) $\mathfrak{g}$-module M_χ generated by one element m_χ satisfying*

(i) $\xi m_\chi = 0 \quad \forall \xi \in \mathfrak{N}$,

(ii) $hm_\chi = \chi(h)m_\chi \quad \forall h \in \mathfrak{h}$, *with the universal property that every other $\mathfrak{g}$-module M generated by one element m satisfying (i) and (ii) is a quotient module of M_χ via a map which sends m_χ to m. The module M_χ is called the* **Verma Module** *for $\mathfrak{g}$ with highest weight χ.*

If P is a proper submodule of M_χ then every weight η of P satisfies $\eta < \chi$. From this it is easy to see that any sum of proper submodules of M_χ is again a proper submodule, so M_χ has a *unique* maximal proper submodule. Equivalently M_χ has a unique irreducible quotient module called L_χ. This module is the unique irreducible $\mathfrak{g}$-module with highest weight χ.

A Verma module M_χ has a filtration by submodules

$$M_\chi = M_{\chi,0} \supseteq M_{\chi,1} \supseteq ... \supseteq M_{\chi,q} = 0$$

such that the quotient modules $M_{\chi,j}/M_{\chi,j+1}$ are all irreducible. This filtration is not necessarily unique but the modules $M_{\chi,j}/M_{\chi,j+1}$ are uniquely determined by M_χ up to isomorphism and change of order. It turns out that these modules are all of the form L_ϕ where $\phi \in \mathfrak{h}_*$ and $\phi \le \chi$, and moreover $\phi + \rho$ lies in the same Weyl group orbit as $\chi + \rho$ where

$$\rho = \frac{1}{2} \sum_{\alpha \in \Sigma^+} \alpha$$

is half the sum of the positive roots of $\mathfrak{g}$. The module L_χ occurs exactly once in the list. An important problem in the study of Verma modules (and hence of all representations of $\mathfrak{g}$) is to determine how many times L_ϕ occurs in the list when $\phi \ne \chi$.

This problem can be rephrased using the **Grothendieck group** of $\mathfrak{g}$-modules. This is the Abelian group generated by isomorphism classes $[M]$ of finitely generated $\mathfrak{g}$-modules M with relations

$$[M_2] = [M_1] + [M_3]$$

for every exact sequence $0 \to M_1 \to M_2 \to M_3 \to 0$ of $\mathfrak{g}$-modules. In the Grothendieck group we can formally write

$$[M_\chi] = \sum_{\phi+\rho \in W(\chi+\rho)} b_{\chi\phi}[L_\phi] \tag{12.1}$$

for some integer coefficients $b_{\chi\phi}$. Our problem then becomes that of determining these coefficients. The coefficient $b_{\chi\chi}$ is always 1, but the other coefficients are more mysterious.

The matrix $(b_{\chi\phi})$ where $\chi + \rho$ and $\phi + \rho$ run over a fixed Weyl group orbit in $\mathfrak{h}^*$ is lower triangular with respect to the partial order $\le$ on $\mathfrak{h}^*$ and has ones on the diagonal. Hence this matrix is invertible. It is more convenient to work with the inverse matrix $(a_{\chi\phi})$ defined by the equation

$$[L_\chi] = \sum_{\phi+\rho \in W(\chi+\rho)} a_{\chi\phi}[M_\phi] \tag{12.2}$$

in the Grothendieck group.

The **Kazhdan–Lusztig conjecture** (see Kazhdan and Lusztig [104], [105], Brylinski and Kashiwara [39], Beilinson and Bernstein [12]) identifies the coefficients $a_{\chi\phi}$ in the special case when $\chi + \rho$ and $\phi + \rho$ lie in the Weyl group orbit of $-\rho$. If w and v lie in the Weyl group W let us write a_{wv} for $a_{\chi\phi}$ and also L_w for L_χ and M_w for M_χ where

$$\chi = w(-\rho) - \rho$$

and

$$\phi = v(-\rho) - \rho.$$

Then the Kazhdan–Lusztig conjecture is concerned with the coefficients a_{wv} satisfying

$$[L_w] = \sum_{v \in W} a_{wv}[M_v] \tag{12.3}$$

in the Grothendieck group. Following Bernstein [15] we shall first identify these coefficients in terms of $\mathcal{D}_X$-modules for a suitable X.

12.2 $\mathcal{D}$-modules over flag manifolds

Recall from Section 12.1 that the Lie algebra $\mathfrak{g}$ of G can be decomposed as

$$\mathfrak{g} = \mathfrak{h} \oplus \left(\bigoplus_{\alpha \in \Sigma} \mathfrak{g}^\alpha \right).$$

Let B be the Borel subgroup of G whose Lie algebra is

$$\mathfrak{b} = \mathfrak{h} \oplus \left(\bigoplus_{\alpha \in \Sigma^+} \mathfrak{g}^\alpha \right).$$

Then

$$X = G/B$$

is the **flag manifold** of G.

Examples 12.2.1. If $G = \mathrm{SL}(n, \mathbb{C})$ as in Example 12.1.1 then we can take B to be the subgroup of $\mathrm{SL}(n, \mathbb{C})$ consisting of upper triangular matrices. Then X can be naturally identified with the space of flags

$$0 \subseteq V_1 \subseteq V_2 \subseteq ... \subseteq V_n = \mathbb{C}^n$$

such that V_j is a j-dimensional subspace of $\mathbb{C}^n$ for each j.

The flag manifold X is a non-singular complex projective variety and G acts transitively on X. Hence if $\mathcal{M}$ is $\mathcal{D}_X$-module then the space $\Gamma(\mathcal{M}) =$

$\mathcal{M}(X)$ of global sections of $\mathcal{M}$ has a natural $\mathfrak{g}$-module structure defined as follows.

Given any $\xi \in \mathfrak{g}$ the infinitesimal action of G on X induces a vector field $x \mapsto \xi_x$ on X. Here ξ_x is the tangent at x to the smooth path

$$t \mapsto \exp(t\xi).x \qquad (t \in \mathbb{R})$$

in X where $\exp \colon \mathfrak{g} \mapsto G$ is the exponential mapping (see e.g. Warner [175]). In local coordinates $z_1, ..., z_n$ we can write

$$\xi_x = a_1(z)\frac{\partial}{\partial z_1} + ... + a_n(z)\frac{\partial}{\partial z_n}.$$

We can define a differential operator D_ξ on X by

$$D_\xi = a_1(z)D_1 + ... + a_n(z)D_n$$

in local coordinates. This gives a Lie algebra homomorphism from $\mathfrak{g}$ to the space $\mathcal{D}_X(X)$ of differential operators on X defined by

$$\xi \mapsto D_\xi.$$

Hence there is a $\mathfrak{g}$-module structure on $\Gamma(\mathcal{M}) = \mathcal{M}(X)$ defined by

$$\xi \cdot \sigma = D_\xi \sigma, \qquad \xi \in \mathfrak{g}, \quad \sigma \in \Gamma(\mathcal{M}).$$

The transitive action of G on $X = G/B$ restricts to an action of B which has finitely many orbits, corresponding to the finitely many double cosets in $B\backslash G/B$. The Bruhat decomposition tells us that these orbits are indexed by the Weyl group W. If $w \in W = N_G(T)/T$ is represented by $\tilde{w} \in N_G(T)$ then the B-orbit X_w of X indexed by w is the B-orbit of the coset $\tilde{w}B$ in X, i.e. the image in X of the double coset $B\tilde{w}B$ in G. The closure $\overline{X}_w$ of any B-orbit X_w in X is a union of B-orbits.

We wish to find $\mathcal{D}_X$-modules λ_w and μ_w supported on $\overline{X}_w$ such that the associated $\mathfrak{g}$-modules $\Gamma(\lambda_w)$ and $\Gamma(\mu_w)$ are naturally isomorphic to L_w and the Verma module M_w. How can we describe these $\mathcal{D}_X$-modules λ_w and μ_w? We can use the Riemann–Hilbert correspondence (Section 11.10) between $\mathcal{D}_X$-modules and intersection sheaf complexes of subvarieties of X.

Consider the intersection sheaf complex $\mathcal{IC}_{\overline{X}_w}^\bullet$ of the irreducible closed subvariety $\overline{X}_w$ of X. By the Riemann–Hilbert correspondence (Theorems 11.10.5 and 11.10.6) there exists a unique irreducible holonomic $\mathcal{D}_X$-module λ_w with regular singularities such that the de Rham complex $DR(\lambda_w)$ of λ_w is isomorphic in $D^b(X)$ to $\mathcal{IC}_{\overline{X}_w}^\bullet$ with a dimension shift.

Let $\mathcal{T}_w^\bullet$ be the sheaf complex on X which is the extension by zero of the trivial sheaf $\mathbb{C}_{X_w}$ on X_w. In other words $\mathcal{T}_w^i$ is zero when i is non-zero, and when i is zero its restriction to X_w is the constant sheaf defined by $\mathbb{C}$ and its stalk at any x not in X_w is zero. There is a $\mathcal{D}_X$-module μ_w on X supported on $\overline{X}_w$ whose de Rham complex $DR(\mu_w)$ is isomorphic in $D^b(X)$ to the sheaf complex $\mathcal{T}_w^\bullet$ with a dimension shift.

Theorem 12.2.2 (Bernstein [15], Beilinson and Bernstein [12], Brylinski and Kashiwara [39]). *The $\mathfrak{g}$-modules $\Gamma(\mu_w)$ and $\Gamma(\lambda_w)$ are isomorphic to M_w and L_w.*

It follows that for suitable integers $d(v, w)$ coefficients a_{wv} defined by equation (12.3) can also be defined by the equation

$$\mathcal{IC}^{\bullet}_{\overline{X}_w} \sim \sum_{v \in w} a_{wv} \mathcal{T}^{\bullet}_v [d(v, w)] \tag{12.4}$$

where $\sim$ denotes the equivalence relation on the free Abelian group of isomorphism classes in $D^b(X)$ of bounded constructible complexes of sheaves on X given by quotienting by the subgroup generated by all elements of the form

$$\mathcal{A}^{\bullet} - \mathcal{B}^{\bullet} + \mathcal{C}^{\bullet}$$

such that there is a distinguished triangle

$$\mathcal{A}^{\bullet} \to \mathcal{B}^{\bullet} \to \mathcal{C}^{\bullet} \to \mathcal{A}^{\bullet}[1].$$

In particular, $\mathcal{A}^{\bullet}[n] \sim (-1)^n \mathcal{A}^{\bullet}$ for any complex $\mathcal{A}^{\bullet}$ so we can replace equation (12.4) by the equation

$$\mathcal{IC}^{\bullet}_{\overline{X}_w} \sim \sum_{v \in W} (-1)^{d(v,w)} a_{wv} \mathcal{T}^{\bullet}_v. \tag{12.5}$$

The **Euler characteristic** of a complex $C^{\bullet}$ of Abelian groups with only finitely many non-zero homology groups is by definition

$$\chi(C^{\bullet}) = \sum_{i \in \mathbb{Z}} (-1)^i \dim H_i(C^{\bullet}).$$

It is easy to check that if

$$0 \to A^{\bullet} \to B^{\bullet} \to C^{\bullet} \to 0$$

is a short exact sequence of complexes then

$$\chi(A^{\bullet}) - \chi(B^{\bullet}) + \chi(C^{\bullet}) = 0.$$

Thus by restricting equation (12.5) to the orbit of X_v and taking Euler characteristics of stalk complexes one finds that

$$a_{wv} = 0 \tag{12.6}$$

unless $X_v \subseteq \overline{X}_w$ and

$$a_{wv} = (-1)^{\dim X_v - \dim X_w} \sum_{i \geq 0} (-1)^i \dim IH^i_{X_v}(\overline{X}_w) \tag{12.7}$$

if $X_v \subseteq \overline{X}_w$ where

$$\dim IH^i_{X_v}(\overline{X}_w)$$

denotes the dimension of the stalk of the $(-i)$th cohomology sheaf of $\mathcal{IC}^\bullet_{\overline{X}_w}$ at any point in X_v.

The question arises whether we can work out the dimensions of these intersection cohomology groups? The answer is that in general we cannot give explicit formulas for their dimensions but we can express them in terms of some polynomials related to Hecke algebras, which can be computed by recursive formulas, given enough time and patience.

The first step is to consider the whole setup in characteristic p where p is a prime number, as in Chapter 10.

12.3   Characteristic p

Let us assume that G is an algebraic group defined over an algebraic number field R (see Springer [165]) and that π is a prime ideal in R such that R/π is isomorphic to the finite field $\mathbb{F}_q$ with $q = p^m$ elements. Let us assume that the reduction of G modulo π is an algebraic group G_q defined over $\mathbb{F}_q$. As in Chapter 10 when we were considering the Weil conjectures we assume that π is not one of finitely many 'bad' primes for G. Then we can assume that the reductions modulo π of the Borel subgroup B, the flag manifold $X = G/B$ and each orbit X_W are respectively a Borel subgroup B_q of G_q, the flag manifold

$$X_q = G_q/B_q$$

and an orbit $(X_w)_q = X_{w,q}$ of B_q on X_q. Then if ℓ is a prime different from p the ℓ-adic intersection cohomology sheaf complex

$$\mathcal{IC}^\bullet_{\overline{X}_{w,q}}$$

and the sheaf complex $\mathcal{T}^\bullet_{w,q}$ given by extending the trivial sheaf complex

$$(\mathbb{Q}_\ell)^\bullet_{X_{w,q}}$$

on $X_{w,q}$ by zero satisfy

$$\mathcal{IC}^\bullet_{w,q} \sim \sum_{v \in W} (-1)^{\dim X_w - \dim X_v} a_{wv} \mathcal{T}^\bullet_{v,q} \qquad (12.8)$$

where the equivalence relation is defined as in Section 12.2.

The Frobenius mapping (Definition 10.2.2) lifts naturally to actions on $\mathcal{IC}^\bullet_{X_{w,q}}$ and on $\mathcal{T}^\bullet_{w,q}$. Let us modify the equivalence relations $\sim$ by considering bounded constructible complexes of ℓ-adic sheaves together with distinguished 'Frobenius' endomorphisms which lift the Frobenius mapping, and quotienting by the subgroup generated by expressions of the form

$$\mathcal{A}^\bullet - \mathcal{B}^\bullet + \mathcal{C}^\bullet$$

for each distinguished triangle

$$\mathcal{A}^{\bullet} \to \mathcal{B}^{\bullet} \to \mathcal{C}^{\bullet} \to \mathcal{A}^{\bullet}[1]$$

which *respects* the Frobenius actions. Let us also write

$$\mathcal{A}^{\bullet} \sim q^{\frac{j}{2}} \mathcal{B}^{\bullet}$$

if $\mathcal{A}^{\bullet}$ is the tensor product of $\mathcal{B}^{\bullet}$ with a 1-dimensional vector space over $\mathbb{Q}_{\ell}$ on which the Frobenius endomorphism acts as multiplication by an algebraic integer of modulus $q^{\frac{j}{2}}$. Then

$$\mathcal{IC}^{\bullet}_{\overline{X}_{w,q}} \sim \sum_{v \in W} p_{wv}(q) \mathcal{T}^{\bullet}_{v,q} \tag{12.9}$$

where $p_{wv}(q)$ is polynomial in $q^{\frac{1}{2}}$ such that

$$p_{ww}(q) = 1 \quad \text{and} \quad p_{wv} = 0$$

if $X_v \not\subseteq \overline{X}_w$ while

$$p_{wv}(q) = \sum_{i \geq 0} (-1)^i q^{i/2} \dim IH^i_{X_v}(\overline{X}_w) \tag{12.10}$$

if $X_v \subseteq \overline{X}_w$. This can be deduced from Riemann hypothesis (10.8) (see Kazhdan and Lusztig [105]).

In particular it follows from the local calculation in Proposition 4.7.2 that if $w \neq v$ then $p_{wv}(q)$ is a polynomial in $q^{\frac{1}{2}}$ of degree less than $\dim X_w - \dim X_v$. In fact

$$IH^i_{X_v}(\overline{X}_w) = 0$$

when i is odd, so $p_{wv}(q)$ is a polynomial in q of degree less than

$$\frac{1}{2}(\dim X_w - \dim X_v).$$

Note that if we formally put $q = 1$ then by comparing equations (12.8) and (12.9) we get

$$p_{wv}(1) = (-1)^{\dim X_w - \dim X_v} a_{wv} \tag{12.11}$$

for all $w, v \in W$.

12.4   Hecke algebras and the Kazhdan–Lusztig polynomials

The **Hecke algebra** H of the Weyl group W of G with parameter q is an algebra over the ring $\mathbb{Z}[q^{\frac{1}{2}}, q^{-\frac{1}{2}}]$ of polynomials with integer coefficients in $q^{\frac{1}{2}}$

and $q^{-\frac{1}{2}}$. As a module over $\mathbb{Z}[q^{\frac{1}{2}}, q^{-\frac{1}{2}}]$ it has a basis consisting of 1 and one element τ_w for each $w \in W$. Its multiplication is uniquely determined by the rules

$$\tau_w \tau_v = \tau_{wv} \tag{12.12}$$

if $w, v \in W$ and $\dim X_{wv} = \dim X_w + \dim X_v$, and

$$(\tau_\sigma + 1)(\tau_\sigma - q) = 0 \tag{12.13}$$

if $\sigma \in W$ acts as a reflection on $\mathfrak{h}$ (see e.g. Bourbaki [27, Ch. IV §2 Ex. 22, 24]).

There is a unique involution $D : H \to H$ satisfying

$$D(q^{\frac{1}{2}}) = q^{-\frac{1}{2}} \tag{12.14}$$

and

$$D(\tau_\sigma + 1) = q^{-1}(\tau_\sigma + 1) \tag{12.15}$$

whenever $\sigma \in W$ is a reflection.

Proposition 12.4.1 (Kazhdan and Lusztig [104, Thm. 1.1]). *For each $w \in W$ there is a unique $C_w \in H$ of the form*

$$C_w = \tau_w + \sum_{v \in W - \{w\}, X_v \subseteq X_w} \widetilde{p}_{wv}(q) \tau_v$$

where $\widetilde{p}_{wv}(q)$ is a polynomial in q of degree less than

$$\frac{1}{2}(\dim X_w - \dim X_v),$$

satisfying

$$DC_w = q^{-\dim X_w} C_w.$$

The polynomials $\widetilde{p}_{wv}(q)$ are called **Kazhdan–Lusztig polynomials**.

Theorem 12.4.2 (Kazhdan and Lusztig [105]). *The polynomials p_{wv} and $\widetilde{p}_{wv}$, defined in (12.9) and Proposition 12.4.1 respectively, coincide.*

Sketch proof Let $\mathbb{C}_{X_q}$ be the trivial sheaf on X_q. If $w \in V$ let $T^{\bullet}_{w,q}$ be the extension by zero of the trivial sheaf on $X_{w,q}$ as in Section 12.3. Consider the set of all formal linear combinations with coefficients in the ring $\mathbb{Z}[q^{\frac{1}{2}}, q^{-\frac{1}{2}}]$ of bounded constructible complexes of ℓ-adic sheaves on X_q with distinguished 'Frobenius' endomorphisms, modulo the equivalence relation defined in Section 12.3. Let $\mathcal{H}$ be the submodule generated by the equivalence classes T_w of $T^{\bullet}_{w,q}$ for $w \in W$ and the equivalence class of $\mathbb{C}_{X_q}$. One can show that $\mathcal{H}$ has a natural $\mathbb{Z}[q^{\frac{1}{2}}, q^{-\frac{1}{2}}]$-algebra structure with $\mathbb{C}_{X_q}$ as multiplicative identity and that there is an isomorphism $\psi : \mathcal{H} \to H$ such that $\psi(T_w) = \tau_w$ for all $w \in W$. Verdier duality (see §7.4) enables one to define an involution Δ of this algebra $\mathcal{H}$ such that

$$\Delta(q^{\frac{1}{2}}) = q^{-\frac{1}{2}}$$

and

$$\Delta(T_{\sigma+1}) = q^{-1}(T_\sigma + 1)$$

if $\sigma \in W$ is a reflection.

It follows from Section 12.3 that the intersection cohomology sheaf $\mathcal{IC}^\bullet_{\overline{X}_{w,q}}$ represents the element

$$T_w + \sum_{v \in W - \{w\}, X_v \subseteq \overline{X}_w} p_{wv}(q) T_v$$

of $\mathcal{H}$. Moreover $\mathcal{IC}^\bullet_{\overline{X}_{w,q}}$ is self-dual with respect to Verdier duality by (7.3) but the Frobenius map is multiplied by the scalar factor

$$q^{-\dim X_w}$$

under this duality. It therefore follows from the uniqueness of the Kazhdan–Lusztig polynomials $\widetilde{p}_{wv}(q)$ that

$$p_{wv}(q) = \widetilde{p}_{wv}(q)$$

for all w and v in W.

The **Kazhdan–Lusztig conjecture** is obtained by combining Theorem 12.4.2 with (12.11) and (12.3). It was proved by Brylinski and Kashiwara and by Beilinson and Bernstein.

Theorem 12.4.3. *(Kazhdan–Lusztig conjecture). The coefficients a_{wv} such that*

$$[L_w] = \sum_{v \in W} a_{wv}[M_v]$$

in the Grothendieck group of $\mathfrak{g}$-modules are given by

$$a_{wv} = (-1)^{\dim X_w - \dim X_v} p_{wv}(1)$$

where $p_{wv}(q)$ are the Kazhdan–Lusztig polynomials.

12.5 Further reading

The original Kazhdan–Lusztig conjecture spawned several similar conjectures about other representation theories:

1. Lusztig [118] for rational representations of finite Chevalley groups;

2. Deodhar, Gabber and Kac [55] for highest-weight modules of Kac–Moody algebras;

3. Lusztig [120] for finite-dimensional representations of quantum enveloping algebras at roots of unity.

Further details can be found in Lusztig's wide-ranging survey [121] of the uses of intersection cohomology in representation theory.

These conjectures are now known to be equivalent (under some mild restrictions); see Kazhdan and Lusztig [106, 107] and [108] for the equivalence of (i) and (ii), and Andersen, Jantzen and Soergel [1] for the equivalence of (ii) and (iii). The Kazhdan–Lusztig conjecture for Kac–moody algebras (ii) has now been solved. There are two cases, for symmetrisable Kac–Moody algebras of positive level see Kashiwara [94], Kashiwara and Tanisaki [98, 101] and Casian [43] and for affine Kac–Moody algebras of negative level see Kashiwara and Tanisaki [99, 100]. It follows that (with certain restrictions) the conjectures (i) and (iii) are also proved. See Kashiwara and Tanisaki [102] and Tanisaki [171] for surveys.

Bibliography

[1] H. H. Andersen, J. C. Jantzen, and W. Soergel. Representations of quantum groups at a pth root of unity and of semisimple groups in characteristic p: independence of p. *Astérisque*, (220):321, 1994.

[2] E. Andronikof. A microlocal version of the Riemann–Hilbert correspondence. *Topol. Methods Nonlinear Anal.*, 4(2):417–425, 1994.

[3] E. Andronikof. Microlocalization of perverse sheaves. *J. Math. Sci.*, 82(6):3754–3758, 1996. Algebra, 3.

[4] M. Atiyah *et al. Representation theory of Lie groups*, volume 34 of *London Mathematical Society Lecture Note Series*. Cambridge University Press, 1979.

[5] W. Baily, Jr. and A. Borel. Compactification of arithmetic quotients of bounded symmetric domains. *Ann. Math. (2)*, 84:442–528, 1966.

[6] M. Banagl, S. Cappell, and J. Shaneson. Computing twisted signatures and L-classes of stratified spaces. *Math. Ann.*, 326(3):589–623, 2003.

[7] G. Barthel, J.-P. Brasselet, K.-H. Fieseler, and L. Kaup. Equivariant intersection cohomology of toric varieties. In *Algebraic geometry: Hirzebruch 70 (Warsaw, 1998)*, volume 241 of *Contemporary Mathematics*, pages 45–68. American Mathematical Society, 1999.

[8] G. Barthel, J.-P. Brasselet, K.-H. Fieseler, and L. Kaup. Combinatorial intersection cohomology for fans. *Tohoku Math. J. (2)*, 54(1):1–41, 2002.

[9] G. Barthel, J.-P. Brasselet, K.-H. Fieseler, and L. Kaup. Combinatorial duality and intersection product: a direct approach. *Tohoku Math. J. (2)*, 57(2):273–292, 2005.

[10] A. Beilinson. How to glue perverse sheaves. In *K-theory, arithmetic and geometry (Moscow, 1984–1986)*, volume 1289 of *Lecture Notes in Mathematics*, pages 42–51. Springer, 1987.

[11] A. Beilinson. On the derived category of perverse sheaves. In *K-theory, arithmetic and geometry (Moscow, 1984–1986)*, volume 1289 of *Lecture Notes in Mathematics*, pages 27–41. Springer, 1987.

[12] A. Beilinson and J. Bernstein. Localisation des *G*-modules. *Acad. Sci. Paris*, 292:15–18, 1981.

[13] A. Beilinson, J. Bernstein, and P. Deligne. Faisceaux pervers. *Astérisque*, 100, 1982. Proc. C.I.R.M. conférence: Analyse et topologie sur les espaces singuliers.

[14] A. Beilinson, V. Ginzburg, and W. Soergel. Koszul duality patterns in representation theory. *J. Am. Math. Soc.*, 9(2):473–527, 1996.

[15] J. Bernstein. Algebraic theory of *D*-modules. Preprint (course given at C.I.R.M. conference on Differential Systems and Singularities, Luminy, 1983. Proceedings published in *Astérisque* 130, Société Mathématique de France, 1985).

[16] J. Bernstein and V. Lunts. *Equivariant sheaves and functors*, volume 1578 of *Lecture Notes in Mathematics*. Springer–Verlag, 1994.

[17] L. Billera and C. Lee. Sufficiency of McMullen's conditions for *f*-vectors of simplicial polytopes. *Bull. Am. Math. Soc. (N.S.)*, 2(1):181–185, 1980.

[18] L. Billera and C. Lee. A proof of the sufficiency of McMullen's conditions for *f*-vectors of simplicial convex polytopes. *J. Combin. Theory Ser. A*, 31(3):237–255, 1981.

[19] J.-E. Björk. *Analytic D-modules and applications*, volume 247 of *Mathematics and its Applications*. Kluwer Academic Publishers, 1993.

[20] A. Borel. *Algebraic D-modules*, volume 2 of *Perspectives in Mathematics*. Academic Press, 1987.

[21] A. Borel and L. Ji. Compactifications of symmetric and locally symmetric spaces. In *Lie theory*, volume 229 of *Progress in Mathematics*, pages 69–137. Birkhäuser, 2005.

[22] A. Borel and J.C. Moore. Homology theory for locally compact spaces. *Mich. Math. J.*, 7:137–159, 1960.

[23] A. Borel and N. R. Wallach. *Continuous cohomology, discrete subgroups, and representations of reductive groups*, volume 94 of *Annals of Mathematics Studies*. Princeton University Press, 1980.

[24] A. Borel *et al. Intersection cohomology*, volume 50 of *Progress in Mathematics*. Birkhäuser, 1984. Notes on the seminar held at the University of Bern, 1983.

[25] W. Borho and R. MacPherson. Représentations des groupes de Weyl et homologie d'intersection pour les variétés nilpotentes. *C. R. Acad. Sci. Paris Sér. I Math.*, 292(15):707–710, 1981.

[26] R. Bott and L.W. Tu. *Differential forms in algebraic topology*, volume 82 of *Graduate Texts in Mathematics*. Springer–Verlag, 1982.

[27] N. Bourbaki. *Groupes et algébres de Lie*. Hermann, 1968.

[28] T. Braden. Perverse sheaves on Grassmannians. *Canad. J. Math.*, 54(3):493–532, 2002.

[29] T. Braden and M. Grinberg. Perverse sheaves on rank stratifications. *Duke Math. J.*, 96(2):317–362, 1999.

[30] T. Braden and R. MacPherson. Intersection homology of toric varieties and a conjecture of Kalai. *Comment. Math. Helv.*, 74(3):442–455, 1999.

[31] T. Braden and R. MacPherson. From moment graphs to intersection cohomology. *Math. Ann.*, 321(3):533–551, 2001.

[32] P. Bressler and V. Lunts. Intersection cohomology on nonrational polytopes. *Compositio Math.*, 135(3):245–278, 2003.

[33] P. Bressler and V. Lunts. Hard Lefschetz theorem and Hodge–Riemann relations for intersection cohomology of nonrational polytopes. *Ind. Univ. Math. J.*, 54(1):263–307, 2005.

[34] P. Bressler, M. Saito, and B. Youssin. Filtered perverse complexes. *Math. Res. Lett.*, 5(1-2):119–136, 1998.

[35] W. Browder. *Surgery on simply-connected manifolds*. Springer–Verlag, 1972.

[36] J.-L. Brylinski. (Co)-homologie d'intersection et faisceaux pervers. *Astérisque*, 92–93:129–158, 1982.

[37] J.-L. Brylinski. Transformations canoniques, dualité projective, théorie de Lefschetz, transformations de Fourier et sommes trigonométriques. *Astérisque*, (140-141):3–134, 251, 1986. Géométrie et analyse microlocales.

[38] J.-L. Brylinski. Equivariant intersection cohomology. In *Kazhdan–Lusztig theory and related topics (Chicago, IL, 1989)*, volume 139 of *Contemporary Mathematics*, pages 5–32. American Mathematical Society, 1992.

[39] J.-L. Brylinski and M. Kashiwara. Kazhdan–Lusztig conjecture and holonomic systems. *Invent. Math.*, 64:387–410, 1981.

[40] S. Cappell and J. Shaneson. Stratifiable maps and topological invariants. *J. Am. Math. Soc.*, 4:521–551, 1991.

[41] H. Cartan and S. Eilenberg. *Homological algebra. Princeton Landmarks in Mathematics.* Princeton University Press, 1999. With an appendix by David A. Buchsbaum, Reprint of the 1956 original.

[42] R. Carter, G. Segal, and I. MacDonald. *Lectures on Lie groups and Lie algebras*, volume 32 of *London Mathematical Society Student Texts.* Cambridge University Press, 1995.

[43] L. Casian. Proof of the Kazhdan–Lusztig conjecture for Kac–Moody algebras (the characters ch $L_{\omega\rho-\rho}$). *Adv. Math.*, 119(2):207–281, 1996.

[44] J. Cheeger. On the spectral geometry of spaces with cone-like singularities. In *Proceedings of the National Academy of Sciences U.S.A.*, volume 76, pages 2103–2106, 1979.

[45] J. Cheeger. On the Hodge thoery of Riemannian pseudomanifolds. In *Proceedings of Symposia in Pure Mathematics*, volume 36, pages 91–146. Amer. Math. Soc., 1980.

[46] J. Cheeger. Hodge theory of complex cones. *Astérisque*, 101–102:118–134, 1983. Proc. C.I.R.M. conférence: Analyse et topologie sur les espaces singuliers (II-III).

[47] J. Cheeger. Spectral geometry of singular Riemannian spaces. *J. Differential Geom.*, 18(4):575–657 (1984), 1983.

[48] J. Cheeger, M. Goresky, and R. MacPherson. L^2-cohomology and intersection homology for singular algebraic varieties. In *Seminar on Differential Geometry*, volume 102 of *Annals of Mathematics Studies*, pages 303–340. Princeton University Press, 1982.

[49] C. Chevalley. Classification des groupes de Lie algébriques. In *Séminaire C. Chevalley, 1956-1958*. Secrétariat Mathématique de l'Ecole Normale Supérieur, 1958.

[50] V. I. Danilov. The geometry of toric varieties. *Russ. Math. Surveys*, 33(2):97–154, 1978.

[51] J. F. Davis and P. Kirk. *Lecture notes in algebraic topology*, volume 35 of *Graduate Studies in Mathematics*. American Mathematical Society, 2001.

[52] P. Deligne. *Equations différentielles a points singuliers réguliers*, volume 163 of *Lecture Notes in Mathematics*. Springer–Verlag, 1970.

[53] P. Deligne. La conjecture de Weil (I). *Publications Math. IHES*, 43, 1974.

[54] P. Deligne. La conjecture de Weil (II). *Publications Math. IHES*, 52, 1980.

[55] V. V. Deodhar, O. Gabber, and V. Kac. Structure of some categories of representations of infinite-dimensional Lie algebras. *Adv. Math.*, 45(1):92–116, 1982.

[56] A. Dimca. *Sheaves in topology.* Universitext. Springer–Verlag, 2004.

[57] A. Dold. *Lectures on algebraic topology.* Springer–Verlag, 1972.

[58] V. Drinfeld. On a conjecture of Kashiwara. *Math. Res. Lett.*, 8:713–728, 2001.

[59] E. Freitag and R. Kiehl. *Étale cohomology and the Weil conjecture*, volume 13 of *Ergebnisse der Mathematik und ihrer Grenzgebiete (3)*. Springer–Verlag, 1988.

[60] G. Friedman. Stratified fibrations and the intersection homology of the regular neighborhoods of bottom strata. *Topology Appl.*, 134(2):69–109, 2003.

[61] W. Fulton. *Introduction to toric varieties*, volume 131 of *Annals of Mathematics Studies.* Princeton University Press, 1993.

[62] S. Gelfand, R. MacPherson, and K. Vilonen. Perverse sheaves and quivers. *Duke Math. J.*, 83:621–643, 1996.

[63] S. Gelfand and Y. Manin. *Homological algebra.* Springer, 1999. Second edition, originally published as *Algebra V*, Vol. 38 of the *Encyclopaedia of Mathematical Sciences*, Springer, 1994.

[64] S. Gelfand and Y. Manin. *Methods of homological algebra. Springer Monographs in Mathematics.* Springer–Verlag, second edition, 2003.

[65] R. Godement. *Topologie algébrique et théorie des faisceaux.* Hermann, 1958.

[66] M. Goresky. Triangulation of stratified objects. In *Proceedings of the American Mathematical Society*, volume 72, pages 193–200, 1978.

[67] M. Goresky, R. Kottwitz, and R. MacPherson. Equivariant cohomology, Koszul duality, and the localization theorem. *Invent. Math.*, 131(1):25–83, 1998.

[68] M. Goresky and R. MacPherson. Intersection homology theory. *Topology*, 19:135–162, 1980.

[69] M. Goresky and R. MacPherson. On the topology of complex algebraic maps. In *Proceedings of International Congress of Algebraic Geometry, La Rabida (1981)*, volume 961 of *Lecture Notes in Mathematics*, pages 119–129. Springer, 1982.

[70] M. Goresky and R. MacPherson. Intersection homology theory II. *Invent. Math.*, 71:77–129, 1983.

[71] M. Goresky and R. MacPherson. Morse theory and intersection homology. *Astérisque*, 101–102:135–192, 1983. Proc. C.I.R.M. conférence: Analyse et topologie sur les espaces singuliers.

[72] M. Goresky and R. MacPherson. Lefschetz fixed point theorem for intersection homology. *Comment. Math. Helvetici*, 60:366–391, 1985.

[73] M. Goresky and R. MacPherson. Simplicial intersection homology. *Invent. Math.*, 84:432–433, 1986.

[74] M. Goresky and R. MacPherson. *Stratified Morse theory*, volume 3. Folge, Bd. 14 of *Ergebnisse der Mathematik und ihrer Grenzgebiete*. Springer–Verlag, 1988.

[75] M. Greenberg. *Lectures on algebraic topology*. Benjamin, 1966.

[76] P. Griffiths. *Topics in transcendental algebraic geometry*, volume 106 of *Annals of Mathematics Studies*. Princeton University Press, 1984.

[77] P. Griffiths and J. Harris. *Principles of algebraic geometry*. *Wiley Classics Library*. John Wiley & Sons, 1994. Reprint of the 1978 original.

[78] R. Hardt. Topological properties of subanalytic sets. *Trans. Am. Math. Soc.*, 211:57–70, 1975.

[79] R. Hartshorne. *Algebraic geometry*, volume 52 of *Graduate Texts in Mathematics*. Springer–Verlag, 1977.

[80] A. Hatcher. *Algebraic topology*. Cambridge University Press, 2002.

[81] P. J. Hilton and U. Stammbach. *A course in homological algebra*, volume 4 of *Graduate Texts in Mathematics*. Springer–Verlag, second edition, 1997.

[82] B. Hughes and S. Weinberger. Surgery and stratified spaces. In *Surveys on surgery theory (Vol. 2)*, volume 149 of *Annals of Mathematics Studies*, pages 319–352. Princeton University Press, 2001.

[83] W. Hurewicz and H. Wallman. *Dimension Theory*. Princeton University Press, 1941.

[84] B. Iversen. *Cohomology of sheaves*. Universitext. Springer–Verlag, 1986.

[85] N. Jacobson. *Lie algebras*. Interscience Publishers, John Wiley & Sons, 1962.

[86] L. Ji. Introduction to symmetric spaces and their compactifications. In *Lie theory*, volume 229 of *Progress in Mathematics*, pages 1–67. Birkhäuser, 2005.

[87] R. Joshua. Equivariant intersection cohomology—a survey. In *Invariant theory (Denton, TX, 1986)*, volume 88 of *Contemperary Mathematics*, pages 25–31. American Mathematical Society, 1989.

[88] V.G. Kac. *Infinite dimensional Lie algebras*. Cambridge Unversity Press, 1985.

[89] G. Kalai. A new basis of polytopes. *J. Combin. Theory Ser. A*, 49(2):191–209, 1988.

[90] K. Karu. Hard Lefschetz theorem for nonrational polytopes. *Invent. Math.*, 157(2):419–447, 2004.

[91] M. Kashiwara. Faisceaux constructibles et systémes holonomes d'equations aux derivées partielles linéaires à points singuliers reguliers. *Séminaire Goulaouic-Schwartz*, 19, 1979–80.

[92] M. Kashiwara. *Systems of microdifferential equations*, volume 34 of *Progress in Mathematics*. Birkhäuser, 1983.

[93] M. Kashiwara. Vanishing cycle sheaves and holonomic systems of differential equations. In *Algebraic geometry (Tokyo/Kyoto, 1982)*, volume 1016 of *Lecture Notes in Mathematics*, pages 134–142. Springer, 1983.

[94] M. Kashiwara. Kazhdan–Lusztig conjecture for a symmetrizable Kac–Moody Lie algebra. In *The Grothendieck Festschrift, Vol. II*, volume 87 of *Progress in Mathematics*, pages 407–433. Birkhäuser, 1990.

[95] M. Kashiwara. *D-modules and microlocal calculus*, volume 217 of *Translations of Mathematical Monographs*. American Mathematical Society, 2003.

[96] M. Kashiwara, T. Kawai, and T. Kimura. *Foundations of algebraic analysis*, volume 37 of *Princeton Mathematical Series*. Princeton Univeristy Press, 1986.

[97] M. Kashiwara and P. Schapira. *Sheaves on manifolds*, volume 292 of *Grundlehren der Mathematischen Wissenschaften*. Springer–Verlag, 1994. Corrected reprint of the 1990 original.

[98] M. Kashiwara and T. Tanisaki. Kazhdan–Lusztig conjecture for symmetrizable Kac–Moody Lie algebra. II. Intersection cohomologies of Schubert varieties. In *Operator algebras, unitary representations, enveloping algebras, and invariant theory (Paris, 1989)*, volume 92 of *Progress in Mathematics*, pages 159–195. Birkhäuser, 1990.

[99] M. Kashiwara and T. Tanisaki. Kazhdan–Lusztig conjecture for affine Lie algebras with negative level. *Duke Math. J.*, 77(1):21–62, 1995.

[100] M. Kashiwara and T. Tanisaki. Kazhdan–Lusztig conjecture for affine Lie algebras with negative level. II. Nonintegral case. *Duke Math. J.*, 84(3):771–813, 1996.

[101] M. Kashiwara and T. Tanisaki. Kazhdan–Lusztig conjecture for symmetrizable Kac–Moody Lie algebras. III. Positive rational case. *Asian J. Math.*, 2(4):779–832, 1998.

[102] M. Kashiwara and T. Tanisaki. On Kazhdan–Lusztig conjectures [translation of Sūgaku **47** (1995), no. 3, 269–285; MR1371157 (97k:17039)]. *Sugaku Expositions*, 11(2):177–195, 1998.

[103] N. Katz. An overview of Deligne's proof of the Riemann hypothesis for varieties over finite fields (Hilbert's problem 8). In *Mathematical developments arising from Hilbert's problems*, volume 28 of *Proceedings of Symposia in Pure Mathematics*, pages 275–306. American Mathematical Society, 1976.

[104] D. Kazhdan and G. Lusztig. Representations of Coxeter groups and Hecke algebras. *Invent. Math.*, 53(2):165–184, 1979.

[105] D. Kazhdan and G. Lusztig. Schubert varieties and Poincaré duality. In *Geometry of the Laplace operator (Proceedings of Symposia in Pure Mathematics, Univ. Hawaii, Honolulu, Hawaii, 1979)*, Proceedings of Symposia in Pure Mathematics, XXXVI, pages 185–203. American Mathematical Society, 1980.

[106] D. Kazhdan and G. Lusztig. Tensor structures arising from affine Lie algebras. I, II. *J. Am. Math. Soc.*, 6(4):905–947, 949–1011, 1993.

[107] D. Kazhdan and G. Lusztig. Tensor structures arising from affine Lie algebras. III. *J. Am. Math. Soc.*, 7(2):335–381, 1994.

[108] D. Kazhdan and G. Lusztig. Tensor structures arising from affine Lie algebras. IV. *J. Am. Math. Soc.*, 7(2):383–453, 1994.

[109] R. Kiehl and R. Weissauer. *Weil conjectures, perverse sheaves and l'adic Fourier transform*, volume 42 of *Ergebnisse der Mathematik und ihrer Grenzgebiete. 3. Folge*. Springer–Verlag, 2001.

[110] H.C. King. Topological invariance of intersection homology without sheaves. *Topology Appl.*, 20:149–160, 1985.

[111] F. Kirwan. Rational intersection homology of quotient varieties. *Invent. Math.*, 86:471–505, 1986.

[112] F. Kirwan. Intersection homology and torus actions. *J. Am. Math. Soc.*, 1(2):385–400, 1988.

[113] S. L. Kleiman. The development of intersection homology theory. In *A century of mathematics in America, Part II*, volume 2 of *History of Mathematics*, pages 543–585. American Mathematical Society, 1989.

[114] S. Łojasiewicz. Triangulation of semi-analytic sets. *Ann. Scuola Norm. Sup. Pisa*, 18(3):449–474, 1964.

[115] S. Łojasiewicz. Sur les ensembles semi-analytiques. In *Actes du Congrès International des Mathématiciens (Nice, 1970), Tome 2*, pages 237–241. Gauthier-Villars, 1971.

[116] E. Looijenga. L^2-cohomology of locally symmetric varieties. *Compositio Math.*, 67(1):3–20, 1988.

[117] E. Looijenga. Cohomology and intersection homology of algebraic varieties. In *Complex Algebraic Geometry at Park City*, IAS / Park City mathematics series 3, pages 221–263. American Mathematical Society, 1997.

[118] G. Lusztig. Some problems in the representation theory of finite Chevalley groups. In *The Santa Cruz Conference on Finite Groups (Univ. California, Santa Cruz, Calif., 1979)*, volume 37 of *Proceedings of Symposia in Pure Mathematics*, pages 313–317. American Mathematical Society, 1980.

[119] G. Lusztig. Character sheaves. I. *Adv. Math.*, 56(3):193–237, 1985.

[120] G. Lusztig. Modular representations and quantum groups. In *Classical groups and related topics (Beijing, 1987)*, volume 82 of *Contemporary Mathematics*, pages 59–77. American Mathematical Society, 1989.

[121] G. Lusztig. Intersection cohomology methods in representation theory. In *Proceedings of the International Congress of Mathematicians (Kyoto, 1990)*, volume I, II, pages 155–174. Mathematical Society of Japan, 1991.

[122] G. Lusztig. Character sheaves and generalizations. Preprint (math.RT/0309134), 2003.

[123] S. Mac Lane. *Categories for the working mathematician*, volume 5 of *Graduate Texts in Mathematics*. Springer–Verlag, second edition, 1998.

[124] R. MacPherson. Global questions in the topology of singular spaces. In *Proceedings of the International Congress of Mathematicians, Warsaw*, volume 1, pages 213–215. Polish Scientific Publishers (PWN), 1983.

[125] R. MacPherson. Intersection homology and perverse sheaves. Colloquium lecture notes distributed by the American Mathematical Society, 1991.

[126] R. MacPherson and K. Vilonen. Elementary construction of perverse sheaves. *Invent. Math.*, 84(2):403–435, 1986.

[127] B. Malgrange. *Équations différentielles à coefficients polynomiaux*, volume 96 of *Progress in Mathematics*. Birkhäuser, 1991.

[128] P. McMullen. The numbers of faces of simplicial polytopes. *Isr. J. Math.*, 9:559–570, 1971.

[129] P. McMullen. On simple polytopes. *Invent. Math.*, 113(2):419–444, 1993.

[130] P. McMullen. Weights on polytopes. *Discrete Comput. Geom.*, 15(4):363–388, 1996.

[131] Z. Mebkhout. Sur le problème de Hilbert–Riemann. *C.R. Acad. Sci.*, 290:415–417, 1980.

[132] Z. Mebkhout. *Le formalisme des six opérations de Grothendieck pour les $\mathcal{D}_X$-modules cohérents*, volume 35 of *Travaux en Cours*. Hermann, 1989.

[133] J.S. Milne. *Étale cohomology*, volume 33 of *Princeton Mathematical Series*. Princeton University Press, 1980.

[134] J. Milnor. *Singular points of complex hypersurfaces*, volume 61 of *Annals of Mathematics Studies*. Princeton University Press, 1968.

[135] J. Milnor. *Morse theory*. Princeton University Press, 1973.

[136] J. Milnor and D. Husemoller. *Symmetric bilinear forms*, volume 73 of *Ergebnisse der Mathematik und ihrer Grenzgebiete*. Springer–Verlag, 1973.

[137] J. Milnor and J. Stasheff. *Characteristic classes*, volume 76 of *Annals of Mathematics Studies*. Princeton University Press, 1974.

[138] O. Neto. A microlocal Riemann–Hilbert correspondence. *Compositio Math.*, 127(3):229–241, 2001.

[139] T. Oda. *Lectures on torus embeddings and applications*, volume 57 of *Tata Institute of Fundamental Research Lectures on Mathematics and Physics*. Springer–Verlag, 1978.

[140] T. Oda. *Introduction to algebraic analysis of complex manifolds*, volume 1 of *Advances in the Study of Pure Mathematics*. North–Holland, 1983.

[141] W. Pardon. Intersection homology Poincaré spaces and the characteristic variety theorem. *Comment. Math. Helv.*, 65(2):198–233, 1990.

[142] M. J. Pflaum. *Analytic and geometric study of stratified spaces*, volume 1768 of *Lecture Notes in Mathematics*. Springer–Verlag, 2001.

[143] F. Pham. *Singularitès des systemes différentiels de Gauss–Manin*, volume 2 of *Progress in Mathematics*. Birkhäuser, 1979.

[144] A. A. Ranicki. *Algebraic L-theory and topological manifolds*, volume 102 of *Cambridge Tracts in Mathematics*. Cambridge University Press, 1992.

[145] A. A. Ranicki, A. J. Casson, D. P. Sullivan, M. A. Armstrong, C. P. Rourke, and G. E. Cooke. *The Hauptvermutung book*, volume 1 of *K-Monographs in Mathematics*. Kluwer Academic Publishers, 1996.

[146] K. Rietsch. An introduction to perverse sheaves. In *Representations of finite dimensional algebras and related topics in Lie theory and geometry*, volume 40 of *Fields Institute Communications*, pages 391–429. American Mathematical Society, 2004.

[147] H. Röhrl. Das Riemann–Hilbertsche Problem der Theorie der linearen Differential-gleichungen. *Math. Annalen*, 133:1–25, 1957.

[148] C. Rourke and B. Sanderson. Homology stratifications and intersection homology. In *Proceedings of the Kirbyfest (Berkeley, CA, 1998)*, volume 2 of *Geometry and Topology Monographs*, pages 455–472. Geometry and Topology Publications, 1999.

[149] Y. Ruan. Stringy geometry and topology of orbifolds. In *Symposium in Honor of C. H. Clemens (Salt Lake City, UT, 2000)*, volume 312 of *Contemporary Mathematics*, pages 187–233. American Mathematical Society, 2002.

[150] M. Saito. Introduction to mixed Hodge modules. *Astérisque*, 179-180:145–162, 1987.

[151] M. Saito. Modules de Hodge polarisables. *Publ. Res. Inst. Math. Sci.*, 24(6):849–995 (1989), 1988.

[152] M. Saito, T. Kawai, and M. Kashiwara. Microfunctions and pseudo-differential equations. In *Lecture Notes in Mathematics*, number 287, pages 265–529. Springer–Verlag, 1973.

[153] L. Saper. L_2-cohomology of algebraic varieties. In *Proceedings of the International Congress of Mathematicians, Vol. I, II (Kyoto, 1990)*, pages 735–746. Mathematical Society of Japan, 1991.

[154] L. Saper. On the cohomology of locally symmetric spaces and of their compactifications. In *Current Developments in Mathematics, 2002*, pages 219–289. International Press, 2003.

[155] L. Saper. L^2-cohomology of locally symmetric spaces, I. Preprint (math.RT/0412353), 2004.

[156] L. Saper. $\mathcal{L}$-modules and the conjecture of Rapoport and Goresky–MacPherson. *Astérisque*, (298):319–334, 2005. Automorphic forms I.

[157] L. Saper and M. Stern. L_2-cohomology of arithmetic varieties. *Ann. Math. (2)*, 132(1):1–69, 1990.

[158] L. Saper and S. Zucker. An introduction to L^2-cohomology. In *Several complex variables and complex geometry, Part 2 (Santa Cruz, CA, 1989)*, volume 52 of *Proceedings of Symposia in Pure Mathematics*, pages 519–534. American Mathematical Society, 1991.

[159] J. Schürmann. *Topology of singular spaces and constructible sheaves*, volume 63 of *Instytut Matematyczny Polskiej Akademii Nauk. Monografie Matematyczne (New Series) [Mathematics Institute of the Polish Academy of Sciences. Mathematical Monographs (New Series)]*. Birkhäuser, 2003.

[160] J.-P. Serre. Faisceaux algébrique cohérents. *Ann. Math.*, 61:197–278, 1955.

[161] J-P. Serre. Zeta and L-functions. In *Arithmetical algebraic geometry*, pages 82–92. Harper and Row, 1965.

[162] P.H. Siegel. Witt spaces: a geometric cycle theory for KO-homology at odd primes. *Am. J. Math.*, 105:1067–1105, 1983.

[163] E. H. Spanier. *Algebraic topology*. Springer–Verlag, 1981. Corrected reprint.

[164] M. Spivak. *A comprehensive introduction to differential geometry*. Publish or Perish, second edition, 1979.

[165] T. A. Springer. *Linear algebraic groups*, volume 9 of *Progress in Mathematics*. Birkhäuser, second edition, 1998.

[166] R. Stanley. The number of faces of a simplicial convex polytope. *Adv. Math.*, 35(3):236–238, 1980.

[167] R. Stanley. Generalized h-vectors, intersection cohomology of toric varieties, and related results. In *Commutative algebra and combinatorics (Kyoto, 1985)*, volume 11 of *Advanced Studies in Pure Mathematics*, pages 187–213. North–Holland, 1987.

[168] R. Stanley. *Combinatorics and commutative algebra*, volume 41 of *Progress in Mathematics*. Birkhäuser, second edition, 1996.

[169] R. Stanley. Recent developments in algebraic combinatorics. *Israel J. Math.*, 143:317–339, 2004.

[170] S. Sternberg. *Lectures on differential geometry*. Prentice-Hall, 1964.

[171] T. Tanisaki. Character formulas of Kazhdan–Lusztig type. In *Representations of finite dimensional algebras and related topics in Lie theory and geometry*, volume 40 of *Fields Institute Communications*, pages 261–276. American Mathematical Society, 2004.

[172] V. A. Timorin. An analogue of the Hodge–Riemann relations for simple convex polyhedra. *Uspekhi Mat. Nauk*, 54(2(326)):113–162, 1999.

[173] L. D. Tráng and Z. Mebkhout. Introduction to linear differential equations. In *Singularities Part 2 (Arcata, Calif. 1981)*, volume 40 of *Proceedings of Symposia in Pure Mathematics*, pages 31–63. American Mathematical Society, 1983.

[174] J.-L. Verdier. Extension of a perverse sheaf over a closed subspace. *Astérisque*, (130):210–217, 1985. Differential systems and singularities (Luminy, 1983).

[175] F. W. Warner. *Foundations of differentiable manifolds and Lie groups*, volume 94 of *Graduate Texts in Mathematics*. Springer–Verlag, 1983. Corrected reprint of the 1971 edition.

[176] I. Waschkies. The stack of microlocal perverse sheaves. *Bull. Soc. Math. Fr.*, 132(3):397–462, 2004.

[177] C. Weibel. *An introduction to homological algebra*, volume 38 of *Cambridge Studies in Advanced Mathematics*. Cambridge University Press, 1994.

[178] A. Weil. Number of solutions of equations over finite fields. *Bull. Am. Math. Soc.*, 55:497–508, 1949.

[179] H. Whitney. Tangents to an analytic variety. *Ann. Math.*, 81:496–549, 1965.

[180] B. Youssin. Witt groups of derived categories. *J. K-theory*, 11(4):373–395, 1997.

[181] G. M. Ziegler. *Lectures on polytopes*, volume 152 of *Graduate Texts in Mathematics*. Springer–Verlag, 1995.

[182] S. Zucker. L_2-cohomology and intersection homology of locally symmetric varieties. In *Singularities, Part 2 (Arcata, Calif., 1981)*, volume 40 of *Proceedings of Symposia in Pure Mathematics*, pages 675–680. American Mathematical Society, 1983.

[183] S. Zucker. L_2-cohomology and intersection homology of locally symmetric varieties. II. *Compositio Math.*, 59(3):339–398, 1986.

[184] S. Zucker. A brief introduction to the L^2-cohomology of locally symmetric varieties. In *Analysis and Geometry 1989*, pages 145–158. Korean Institute of Technology, 1989.

[185] S. Zucker. L^2-cohomology and intersection homology of locally symmetric varieties. III. *Astérisque*, (179-180):11, 245–278, 1989. Actes du Colloque de Théorie de Hodge (Luminy, 1987).

Index

*For Product Safety Concerns and Information please contact
our EU representative GPSR@taylorandfrancis.com Taylor & Francis
Verlag GmbH, Kaufingerstraße 24, 80331 München, Germany*

T - #0026 - 160425 - C0 - 234/156/13 [15] - CB - 9781584881841 - Gloss Lamination